Report Graphics

A Handbook for Writing the Design Report

Report Graphics

A Handbook for Writing the Design Report

Richard L. Austin ASLA

VNR VAN NOSTRAND REINHOLD COMPANY
NEW YORK CINCINNATI TORONTO LONDON MELBOURNE

Library of Congress Catalog Card Number: 82-23844
ISBN: 0-442-20886-3

Manufactured in the United States of America

Published by Van Nostrand Reinhold Company Inc.
135 West 50th Street,
New York, N.Y. 10020

Van Nostrand Reinhold Reinhold Company Limited
Molly Millars Lane
Wokingham, Berkshire, England

Van Nostrand Reinhold
480 Latrobe Street
Melbourne, Victoria, 3000, Australia

Macmillan of Canada
Division of Gage Publishing Limited
164 Commander Boulevard
Agin Court, Ontario M15 3C7, Canada

15 14 13 12 11 10 9 8 7 6 5 4 3 2 1

Library of Congress Cataloging in Publication Data

Austin, Richard L.
Report graphics.

Bibliography: p.
Includes index.
1. Engineering graphics. I. Title
T353.A87 1983 604.2'4 82-23844
ISBN 0-442-20886-3

Preface

Clear, concise communication is the principal foundation upon which a successful design consultation organization is structured. Concept sketches, illustration art, working drawings, and construction specifications are all part of the important repertoire of communication elements the designer, as the transmitter of technical information, values. Each element independently conveys a vital ingredient that influences the environments created by the design team. Although an individual communication technique has an effect on the very nature of the finished product, the greatest impact is felt when all the ingredients are combined into a refined *design report*.

The communication of design ideas is very difficult, even when there is only one designer reaching out to only one client. A set of working drawings or a detailed specification is far too complicated for the untrained eye to understand, and perspective sketches or three-dimensional scale models may lack the "selling" ability for finalizing a concept.

The *design report* is more than the mere end result of a planning project or even an in-house medium for graphic displays. It is a valuable source of information complied for both client and designer, and effective research efforts are needed for its writing and production. A poorly written and produced document consumes organizational energies and very often creates dissatisfaction among client groups. And, since bad news travels quickly in this highly competitive industry, the future survival of the professional design firm may well depend upon the efforts exerted in creating this product.

Although we are solidly entrenched in the age of microprocessors and multi-image printouts, the design report remains the primary vehicle for architectural, engineering, and planning services communications. It acts as a sales representative, an in-house staff communicator, and even a public relations tool for the promotion of service programs. With such a heavy burden of responsibility falling upon a report document, the successful practitioner must use valuable company resources on its production.

The following material has been developed to assist both the student and the professional producing a design report. It is a guide that will assist designers in developing a good writing style that will help communicate technical data to a nontechnical readership. It presents techniques for balancing descriptive text with informational graphics for more effective communication. It does not restrict creativity—it expands it.

Acknowledgments

I wish to take this opportunity to express my appreciation to the following organizations for their assistance with the preparation of this book. To the Gunning-Mueller Clear Writing Institute, Inc. for the use of the Gunning Fog Readability Index. To Grid Publishing, Inc. for the use of their checklist questions. To Letraset USA, Inc. for the use of their sample letters and screens. And, to the following agencies and firms for the use of their design reports:

Johnson, Johnson, and Roy, Inc.
Ann Arbor, Michigan

Kansas State University
Manhattan, Kansas

Myrick, Newman, Dahlberg and Partners
Dallas, Texas

The Department of Energy
Washington, D.C.

The President's Committee On Employent of the Handicapped
Washington, D.C.

The Superintendent of Documents
Washington, D.C.

The Texas Parks and Wildlife Department
Austin, Texas

The University of Nebraska
Lincoln, Nebraska

The US Fish and Wildlife Service
Washington, D.C.

U.S. Department of the Interior
Washington, D.C.

Contents

Preface / v

Acknowledgments / vii

1—Development / 1

2—Structure / 21

3—The A/V Report / 131

Appendix 1—Folding / 139

Appendix 2—Binding / 143

Appendix 3—Paper and Reduction / 145

Appendix 4—Slide Copy Table / 147

Appendix 5—Tables and Graphs / 149

Glossary / 155

References / 157

Index / 159

Report Graphics

A Handbook for Writing the Design Report

1. Development

THE WRITING PROCESS

The actual writing of a design report is probably the most difficult task a designer will undertake during a consultation program. Because designers are trained and conditioned to use illustrative graphics to express their ideas, the written thought is considered to be an almost insurmountable obstacle by some. Short narrative descriptions often prove to be more troublesome and confounding than lengthy and detailed sheets of construction drawings.

Writing involves the arrangement of technical language and sometimes complicated terminologies into an effective communication tool. It requires a basic knowledge of word usage, sentence construction, and paragraph arrangement, all of which go into a final product that someone other than the author will read, understand, and base future actions upon.

Writing and editing the report document are an important part of the overall design program and should therefore be developed by a thorough and comprehensive process. The readership and the information that group needs should be identified, narrative outlines should be organized, and supportive productions systems should be geared to the completion of this valuable product.

Identifying Objectives

Every report should have a reason to exist. Will it communicate a new or revised idea; persuade someone to complete a specific task; or will it educate a group of people before a major planning program is begun? Regardless of its goal, the design report must be prepared for a specific reason: without one, too much time, effort, and expense may be wasted on its production.

To properly identify the objectives of a report, the report writer must first take a close look at the target readership, or audience, and determine what kind of information the group being addressed needs. This group of people will ultimately determine the success or failure of the production efforts by either accepting or rejecting the composition. The types of readers may vary from report to report, but will inevitably include the technical audience, the management audience, or the general audience. (Figure 1-1)

The *technical audience* is comprised of other professionals who work in the same field as the production team. They usually have a detailed knowledge of the important issues addressed in the report and are apt to be the most critical of any reviewers. This group will "pick" at both the graphic styles in the product as well as the general design program represented in the report. Simple "eyewash" or diversionary "smoke screen" techniques for presenting information will not deflect their resolve to find fault with the work. They are persistent, unforgiving, and the most difficult to please.

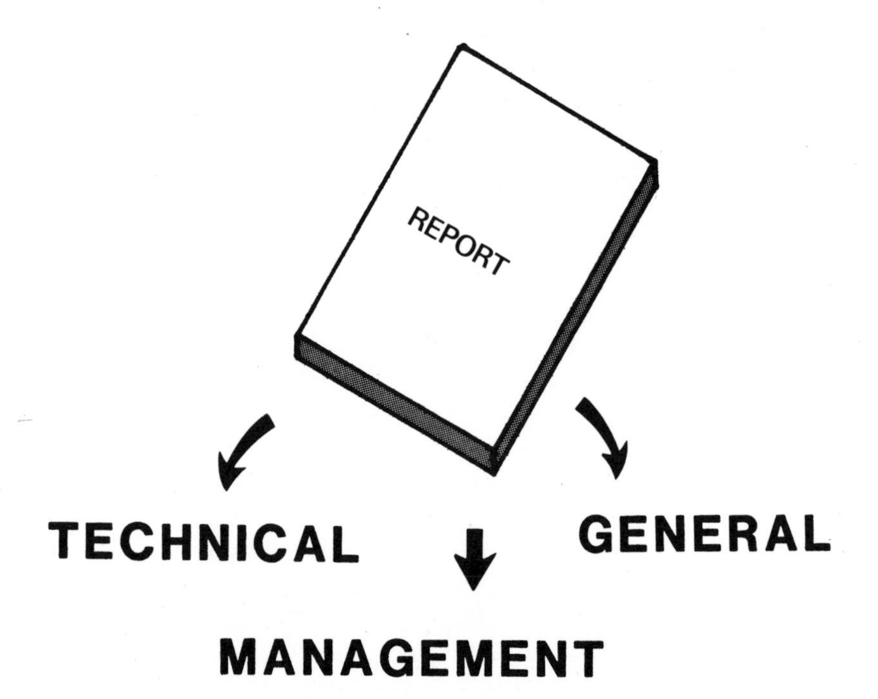

Figure 1-1. The design report must be written to serve a specific audience. It is difficult for one report to adequately serve all types of audiences.

The *management audience*, on the other hand, is often concerned less with technical data and more with goal attainment and cost analysis. These are the individuals who may eventually make the final decision on whether to cancel the project or proceed with its development. Technical data can be used in the composition, but placement should be strategic and should support the program's goal. Criticism may be difficult to come by, because this group will often remain silent after reading the report. They give careful consideration to what is presented, then react at a later time.

The *general audience* is comprised of individuals who have a common interest in the report material but will not have a major impact on the success or failure of the proposed program. They commonly need more graphics or "visuals" to better understand the written material, and technical jargon should be avoided whenever possible. The overall length of the report may be shorter than for other audiences, with only the design highlights or final goal statements used as a main theme. Criticism will normally be good but low-key, unless the nature of the proposed program is politically sensitive.

It should be a goal of the design report writer to reach as much of the audience as possible—at least 80 percent—as traffic signs are designed to reach the total audience to avoid accidents. For the design report to be successful, it is important to remember the following: (Figure 1-2)

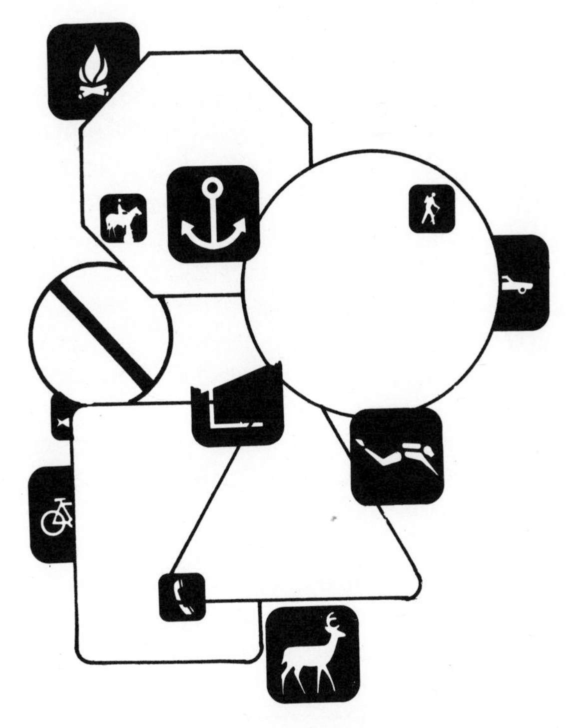

Figure 1-2. Traffic signs and graphic symbols must be designed to reach 100 percent of the audience they serve. Report documents should be written to reach at least 80 percent.

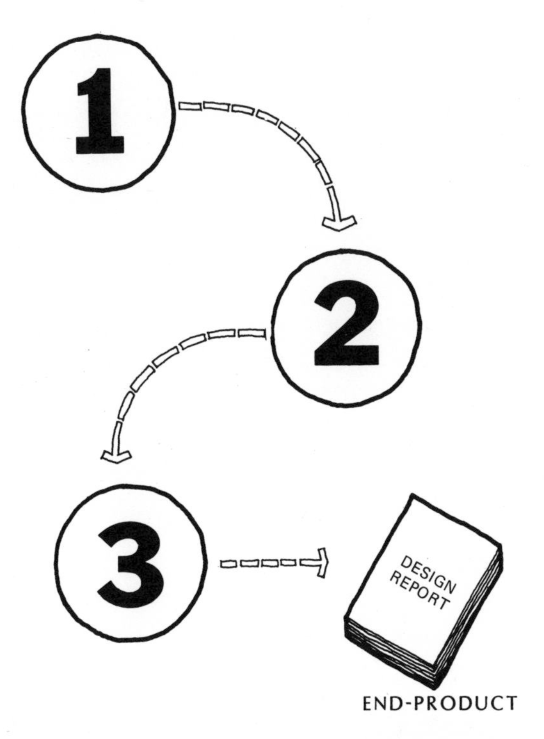

Figure 1-3. Design report writing can be as easy as one, two, three if the author will first consider the end product. First determine the type of information ncessary for the target audience, and then begin to work on the format and information structure.

1. The depth of perception of the audience must be considered. To what degree will the data be analyzed?
2. Will certain color combinations for paper and/or covers be more readily accepted by some audience?
3. Will one type style be more effective than another? What type size should it be, how should it be used, and where should it be used?

Identifying the audience, and its information needs will help prevent "information overkill" and may well determine the success or failure of the project. (Figure 1-3)

Developing the Outline

There are two important questions to ask before beginning a design report outline. First, how is the report to be used? Is it for communicating project feasibility; is it for presenting the resource analysis; or is it the final presentation of overall program concepts? Second, who will read the report? Is the audience technical, managerial, or general. The answers to these questions will control the final framework of the document—they are very important considerations!

The first step in the writing process is the development of an *outline*. What the report must say, how it is to say it, and where it is to be said are critical factors if proper communication is to be maintained. Whether preliminary or formal, the outline serves three basic functions. First, it helps to arrange the factual information that is to be presented in the document. Second, it will assist the writer to plan the information in a logical order of priority. Third, it can be used to guide the writer in adding or cutting before the final typing is begun.

The symbols for the outline will be governed by the audience that will read the report. Roman numerals are best for managerial or general audiences, while decimal symbols are better for the technical group.

ROMAN NUMERALS	DECIMAL SYMBOLS
I.	1.0
A.	1.1
1.	1.11
2.	1.12
B.	1.2
II.	2.0
A.	2.1
B.	2.2
1.	2.21
a.	2.211
b.	2.212
2.	2.22

Both provide a flow to the material when the "I" or the "1.0" is used as the major headings.

An effective technique for outlining a report is the use of index cards. With a set of 5 × 8 cards, color coded if necessary, the title and major sections can be posted on a large board for more effective organization. Then the subsections can be arranged in a logical order to accommodate the audience. As information is developed, it can be placed on a card and added to the outline and moved around the board for greater impact. Once the "card board" is developed, the location of illustrations to support the text can be identified and roughed in on the back of the card. Topic sentences, or even paragraphs, can be written on the face of the card to act as a preliminary draft. This method prevents the writing of a lengthy outline that is always subject to change, thus eliminating or avoiding wasted efforts by the production team.

The following is an example of a procedure that may be followed to complete a report outline by the "card board" method.

Step 1. *Major headings.* Identify the most important categories of information needed in the design report. These will serve as the preliminary chapters of the document and will guide the research and writing of the final composition. (Figure 1-4)

Step 2. *Subheadings.* Within each major heading, make a list of important information that should appear in each chapter. This list will begin to qualify the composition and target specific data for the type of audience the report will serve. (Figure 1-5)

Step 3. *Sections.* This information determines the final flow of the material. The writer can use this to easily adjust the impact of the report and guide future actions by the client. Short, topical sentences that will relate to future paragraphs should be written on each card. (Figure 1-6)

DEVELOPMENT
+ THE WRITING PROCESS
+ WRITING STYLE

PART I

STRUCTURE
+ THE BASIC COMPONENTS
+ PAGE COMPOSITION
+ THE SUPPORTIVE MATERIALS

PART II

THE A/V REPORT
+ ORGANIZATION
+ 35-MM SLIDES
+ EXHIBITS
+ MOTION PICTURES

PART III

Figure 1-4. The first step in the report writing process is to identify the *major headings* of the composition. These components will act as chapters or parts and will guide the development of additional information in the form of subheadings.

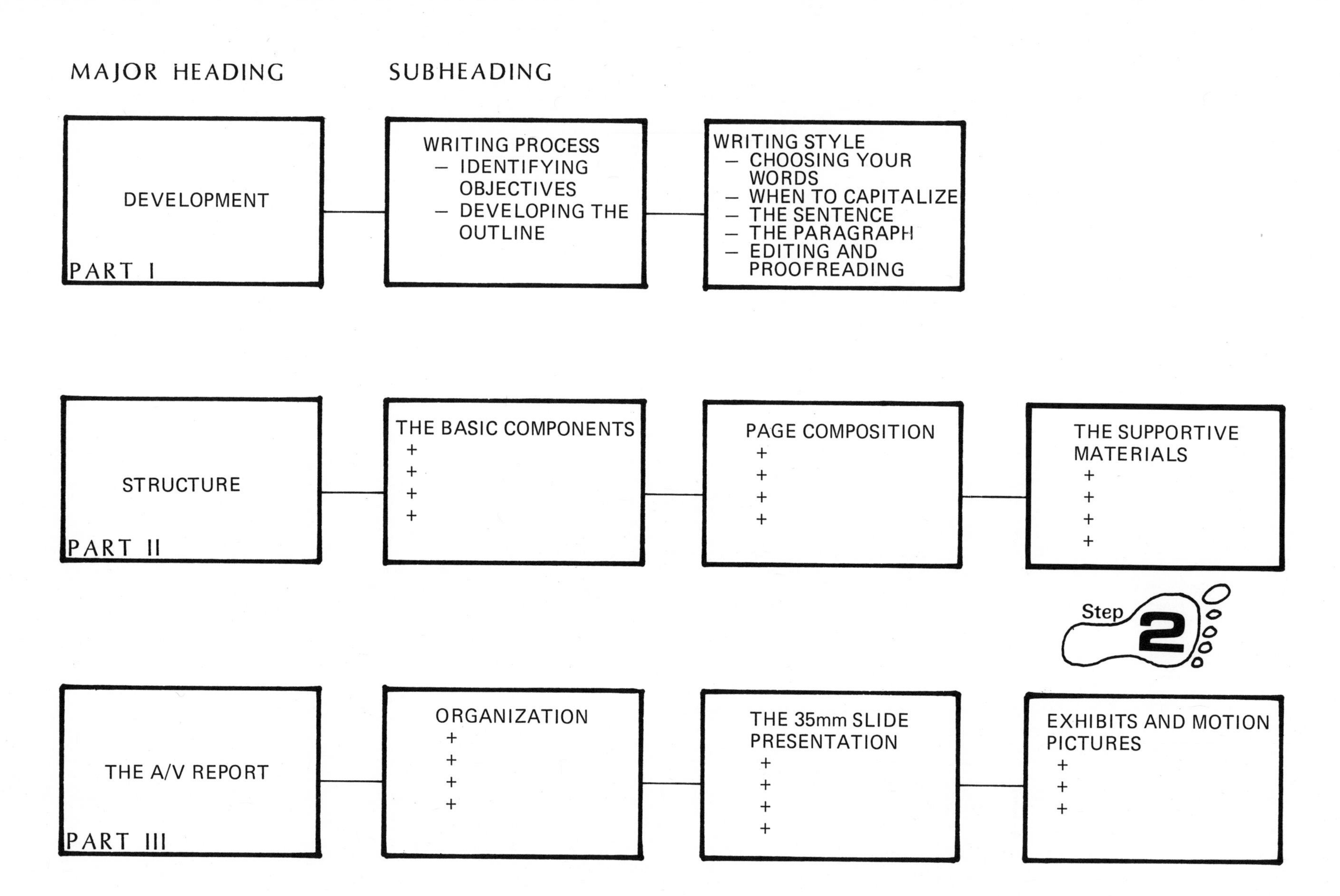

Figure 1-5. The next step in the process is the arrangement of the *subheadings*. Each subject area developed from the previous step should be written on a separate card and then arranged according to priority.

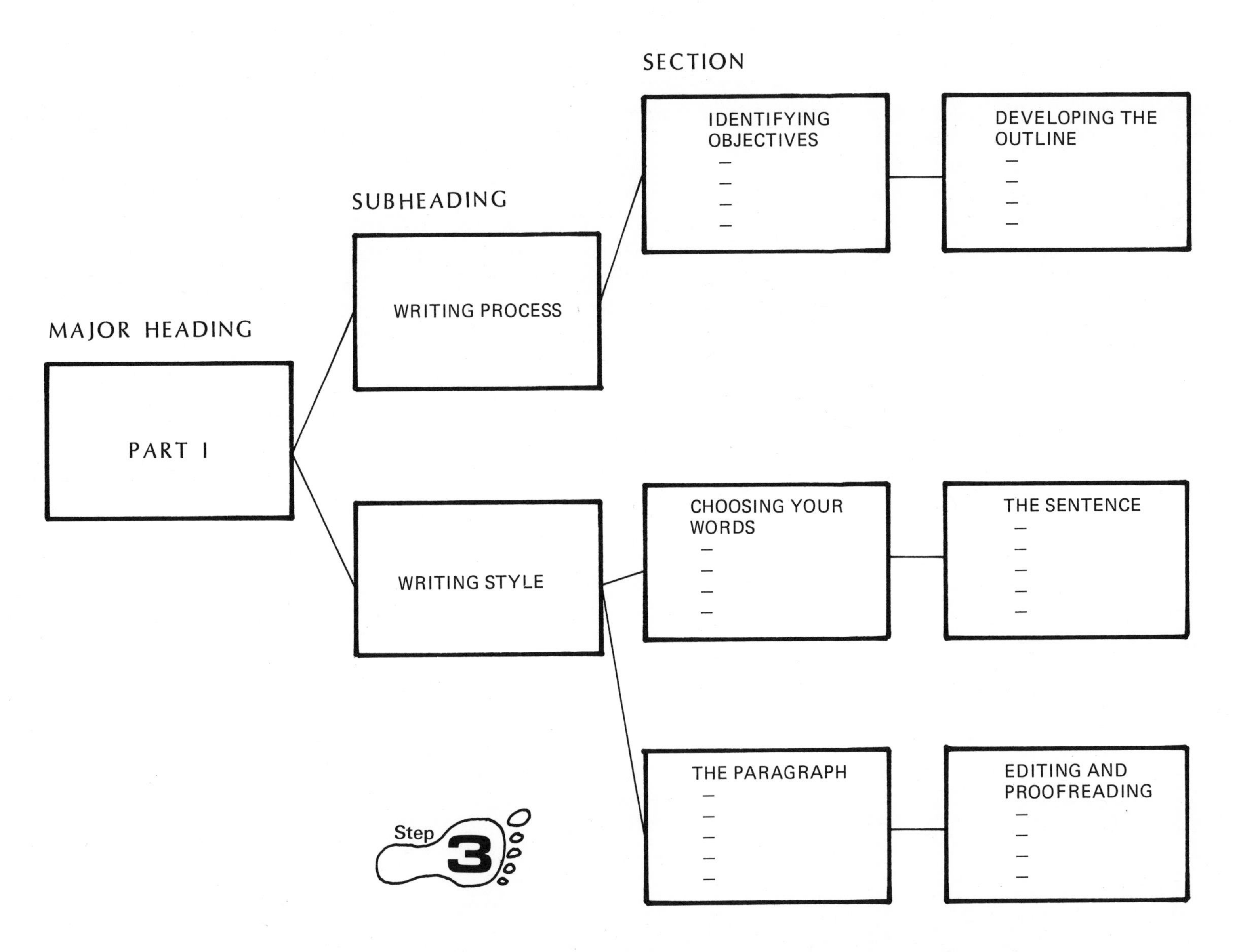

Figure 1-6. From the subheadings, develop the *sections* of information needed in the composition. Begin to expand this data with abbreviated notes on important material.

Step 4. *Subsections.* This information will become the subject areas for each paragraph or groups of paragraphs. Preliminary drafts can be written on the cards before typing, reducing the amount of time needed for the final writing. At this stage of the outlining process, the writer should also be able to determine where and to what extent graphics are needed to support the composition. (Figure 1-7, Figure 1-8.)

WRITING STYLE

Choosing Your Words

The words used for a design report will essentially determine the success or failure of the final composition. Incorrect or improperly used terms divert the reader's attention from the important design issues within the report. By rejecting the written report, the audience may also reject the project. Some important "do nots" for word usage are:

1. Do not use slang—it may be offensive to some readers.
2. Do not use profanity.
3. Do not use professional jargon when addressing a general or management audience.
4. Do not use personal pronouns in a formal report.
5. Do not allow a word to have more than one meaning in a report.
6. Do not invent new words.
7. Do not use shorthand forms of words.
8. Do not use foreign words—use their English equivalents.
9. Do not use contractions.

The most common mistakes are made when writers use words that sound alike but are actually different. One or two leters may separate the proper from the improper, and it is important to pay careful attention to usage of these words. Some of the more troublesome are as follows:

1. Accept/except
   - accept: to receive (a thing offered); admit and agree to
   - except: to take or leave out (anything); to exclude
2. Advise/advice
   - advise: to give information or advice
   - advice: the information given
3. Affect/effect
   - affect: to produce an influence on or alteration in
   - effect: to bring about; to accomplish
4. Adapt/adopt
   - adapt: to make suitable
   - adopt: to accept and apply into practice
5. Among/between
   - among: in or amidst; surrounded by
   - between: in the time, space, or interval that separates

The examples of similar-sounding words and common errors were provided by the *Killeen Daily Herald,* Kileen, Texas. The definitions for the similar-sounding words were adapted from *Webster's New Collegiate Dictionary,* G.&C. Merriam Company, Springfield, Massachusetts, 1977.

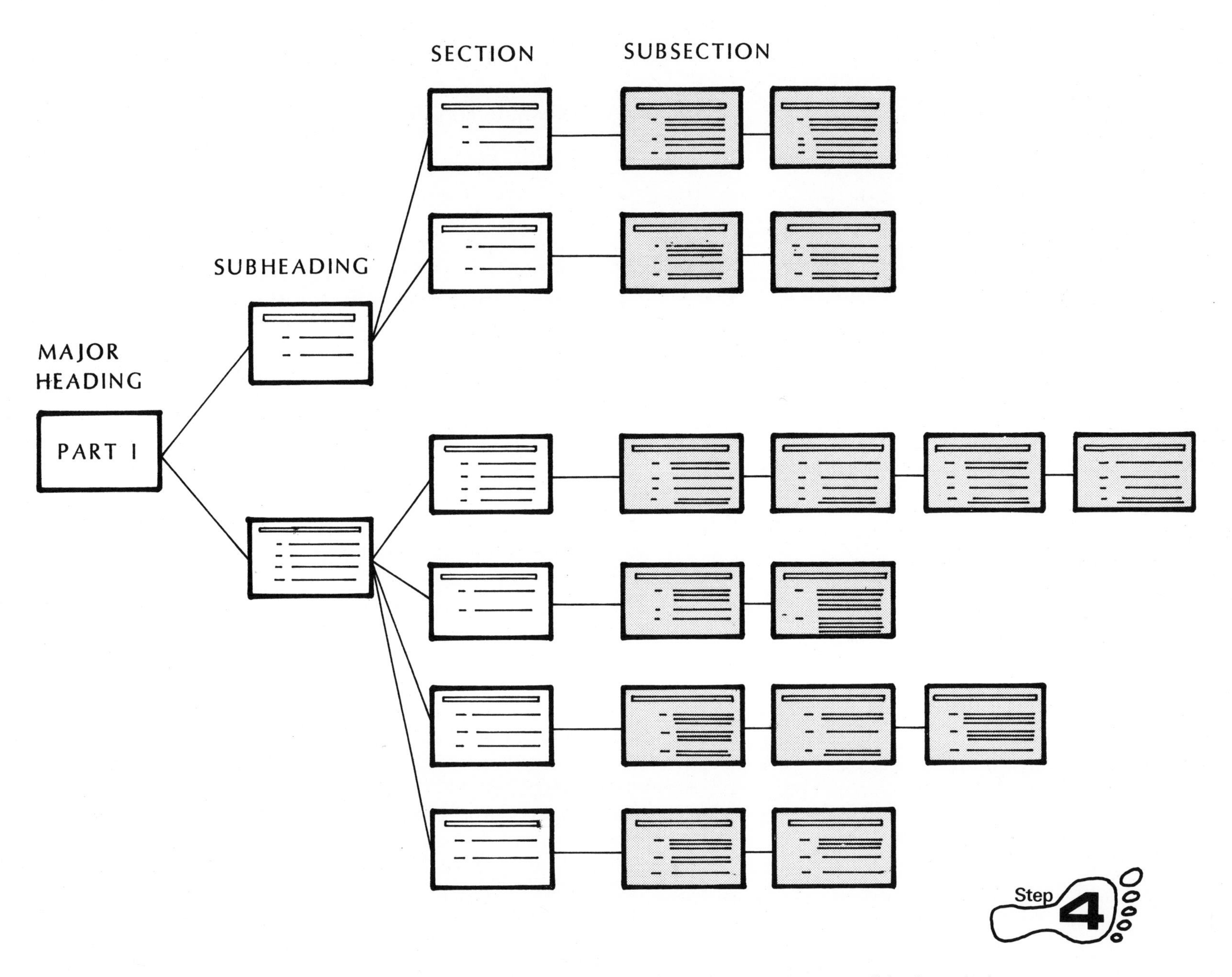

Figure 1-7. The final step in the process is the writing of short sentences or lists of needed information. This is done on the *subsection* cards, where the first draft of the report begins.

6. Assure/insure
   - assure: to make safe; to reassure; to convince
   - insure: to make certain; to give, take, or procure insurance on or for
7. Compliment/complement
   - compliment: an expression of admiration, esteem, or respect
   - complement: that which completes or fills up
8. Device/devise
   - device: that which is formed by design
   - devise: to form in the mind by ideas
9. Farther/further
   - farther: at a great distance
   - further: going beyond; additional
10. Foreword/forward
   - foreword: a word beforehand; a preface
   - forward: advanced, in front; to transmit
11. Later/latter
   - later: after a certain time
   - latter: more recent; the thing mentioned last
12. Loose/lose
   - loose: unattached, disconnected
   - lose: to miss from one's possession
13. Principal/principle
   - principal: a head, leader, chief; main
   - principle: a source of origin; a law, doctrine, or assumption
14. Respectful/respective
   - respectful: showing respect or deference
   - respective: particular; separate
15. Stable/staple
   - stable: steady in purpose; fixed
   - staple: a place of supply; chief constituent of anything
16. Stationary/stationery
   - stationary: fixed in place
   - stationery: papers, pens, blank books, etc.

Some of the other errors commonly made in report writing are:

1. *Plus* used for the conjunction *and:*
   The quarterback gained eight yards, *plus* (instead of *and*) he picked up the first down.
2. *Like* used for *as* or *as if:*
   like (instead of *as*) I was saying. . . .
   He acts *like* (instead of *as if*) he. . . .
3. Adjectives used as verbs:
   Watch *close* (instead of *closely*) as I. . . .
4. *Hopefully* used to mean *I hope that, let us hope that,* or *we hope that:*
   Hopefully, we will (instead of *we hope that* we will) finish. . . .
5. *Less* used when *fewer* (things that can be counted) is needed.
   There are *less* (instead of *fewer*) trees. . . .
6. *Type* (of), *kind* (of), *situation,* and *activity* added without conveying any real information:
   A very open *type of* design. . . .
   A construction-*type situation*. . . .
   Design *activity*. . . .

I. Development

   A. The Writing Process
      1. Identifying objectives
         a. audience identification
         b. when/where to begin
      2. Developing the outline
         a. card board technique
         b. sample outline

   B. Writing Style
      1. Choosing your words
         a. "do nots" of word usage
         b. common mistakes of sound-alike words
         c. errors of the written word
         d. when to capitalize
      2. The sentence
         a. clauses
         b. the types of sentences
      3. The paragraph
         a. length
         b. unity
         c. cohesiveness
      4. Editing and proofreading
         a. readability
         b. checklist

Figure 1-8. This is the finished outline developed by the cardboard process presented on the preceding pages. This method allows the information to flow naturally without making excessive changes or rewrites.

When to Capitalize

The basic rule for capitalizing names of specific persons, places, and things applies to the major- ity of report communications, but when in doubt, consult a current dictionary. Additional words that should be capitalized are: (Figure 1-9)

1. Compass points used to identify specific geographic areas; e.g., Pacific Northwest.
2. The names of races and religions.
3. Political divisions (continents, countries, regions, states, towns) and topographical names (bodies of water, mountains, etc.)
4. Civil, military, and religious titles and offices.
5. The names of organizations, internal units of organizations, and professional positions.
6. Names that identify items or places; e.g., Exhibit A, Form 990.
7. The first letter of all nouns, pronouns, adjectives, verbs, adverbs, and subordinate conjunctions in titles of books, periodicals, newspapers, articles, pamphlets, brochures, etc.
8. The days of the week and months of the year.
9. The brand names, trade names, and trademarks of products.
10. Important words in a list, outline, quotation, or sentence, e.g., Berms, Planted Stock.

The Sentence

The written sentence of a design report is comprised of clauses. A *clause* is a group of related words with a subject and a verb. When the clause expresses a complete thought, it is called an *independent clause;* when it does not express a complete thought, it is called a *dependent clause.*

Because they were the best qualified firm—(dependent clause)—they were awarded the contract—*(independent clause).*

There are four basic types of sentences that may be used in the design report: the simple sentence, the compound sentence, the complex sentence, and the compound-complex sentence. The *simple* sentence contains one independent clause. The *compound* sentence has two or more independent clauses but does not have any dependent clauses. A *complex* sentence has one independent clause and one or more dependent clauses. The *compound-complex* sentence is comprised of two or more independent clauses and at least one dependent clause.

The Paragraph

The design report writer combines words into sentences, sentences into paragraphs, and paragraphs into a comprehensive presentation of important information. The specific combination of these facts is critical if the final report is to be successful. The following criteria should be considered when developing and arranging the written paragraph:

1. *Length.* The maximum length should be 10 to 12 printed lines. The shorter paragraph is easier to read and understand and will help reduce the complexity of the report. (Figures 1-10 and 1-11)
2. *Unity.* One paragraph should present one thought or idea. If there are different parts to an idea or thought, use a section heading and present each part in a separate paragraph.
3. *Cohesiveness.* Maintain a logical order. If the thoughts within a paragraph ramble, so will the minds of the readers.

?

WhEn tO

CapiTaliZe ?

?

always

Capitalize

specific....

*** Persons**

*** Places**

*** Things**

Figure 1.9.

Figure 1-10. The paragraph on the left is lengthy and cumbersome to read. The same information can be presented in shorter paragraphs, even if smaller type is required.

Existing Interpretive Facilities

Currently, visitors to the Frijoles Canyon headquarters area receive information and folders at the visitor center, where they can choose to purchase sales publications, slides, and postcards, see a 10-minute audiovisual (AV) orientation program, visit the museum, or pick up a trail guide to visit the excavated sites of Tyuonyi ruin, the cave dwellings, and Long House. Visitors can continue on to Ceremonial Cave. During the summer months, guided tours are offered to the major ruins area 6 times a day. People who do not wish to take the long ruins trail can visit the excavated Rainbow House a short distance from the visitor center. Those wishing to view the falls and the Rio Grande can hike downcanyon. Hikers into the backcountry, with 65 miles of trails, leave from the same building to see other features such as Stone Lions, Yapashi, San Miguel, and Painted Cave, none of which are excavated. No major changes in visitor flow in and out of the visitor center are anticipated.

Present exhibits are more than 30 years old and outmoded both in content and presentation. A number of specimens are quite good, but many are on loan from the Museum of New Mexico. A valuable collection of artwork by Pablita Velarde and Helmut Naumer is in excellent condition and park-owned. Some of this material can be incorporated into proposed new exhibits, be displayed in public rooms, or be reproduced in publications. In any case, these items should be preserved when new exhibits are installed.

The AV orientation program deals with the trail to the ruins, Ceremonial Cave, backcountry attractions, natural history, seasonal changes, and the detached section containing Tsankawi ruin. Changes will have to be made for the proposed addition of the Canada de Cochiti Grant.

Present tour booklets and the area sales handbook must be updated along with the proposed new interpretive exhibits and other presentations, to avoid contradictions in the interpretive program.

The detached Tsankawi section, with a contact station, offers the visitor a self-guiding trail to the unexcavated site of Tsankawi ruin and will require only editorial changes until excavated.

The current interpretive story presented in sales publications, the museum, and the AV program is based on the obsolete ideas of the 1930's, which draw the culture and ancestry of the prehistoric occupants of Bandelier from the Four Corners area. New information derived from research projects provides pertinent data for revising the story, as well as for more closely relating the people of Bandelier to surviving neighboring pueblos.

visitor with some knowledge of household contents in late prehistoric times. These items can be furnished by the area staff. A skylight in the roof hatchway and a transparent plastic sheet over the doorway interior would protect the room, while keeping it visible to almost all visitors in the near vicinity. One of the rooms of the reconstructed talus house should be left unfinished to illustrate prehistoric building techniques.

Avoid lengthy demonstrations of modern Indian crafts, because visitors do not have the time to see them through. These are better handled by AV at campfire programs and so forth. However, demonstrations should be conducted so long as they can be accomplished in a relatively short time. Visitors in the Frijoles Canyon area can be referred to the Frijoles Canyon Curio Shop and to the arts and crafts shops at nearby pueblos. Competition in sales should be avoided.

VISITOR CENTERS AND CONTACT STATIONS

No new construction will take place in Frijoles Canyon. The existing parking area is sufficient for off-season visitation. During periods of heavy visitor use (May through September), there will be no private vehicular access to the canyon. Instead, visitors will park their vehicles in the new Frijoles Mesa parking area and enter the new Frijoles Mesa contact station, where they will be oriented to the area. A public transit system will take them into Frijoles Canyon utilizing the existing road. If the visitor wishes to walk into the canyon, a trail will be provided that closely follows one of the prehistoric routes, thus letting him experience the transportation of the prehistoric inhabitants.

The open exhibit of religious aspects of the area should be avoided, except in a very general sense, because ritual objects and customs of late prehistoric times resemble those in use today and would offend not only the neighboring Cochiti and San Ildefonso but nearly all pueblos and most other Indian groups.

Although the area collections and exhibitable specimens are small, loan material can be obtained, if needed, until proposed excavations in Tsankawi Pueblo and in the Genizaro or Spanish sites provide additional material.

Frijoles Mesa Contact Station
This new contact station will serve as the primary facility to orient the visitor to the resources of the monument. Emphasis will be placed on Frijoles Canyon and the wilderness area. Brief coverage of the Tsankawi and

Figure 1-11. This is an example of various paragraph lengths used in a design report. They provide the audience with a pleasant reading surface and help to increase the overall acceptance of the information.

Editing and Proofreading

The editing of the report manuscript is one of the most important steps in the process. It is at this point that the thoughts expressed and words written by the writer begin to take their final form. With proper editing and proofreading, reader interest can be heightened and mistakes in information or flow can be corrected.

These tasks should be completed by someone other than the author. Writers sometimes "stand too close to the forest to see the trees," and a second opinion is very important. Reading the material aloud will also help the editor check for clarity, organization of facts, and the unity of thought.

Standard marks have been adopted by the printing and publishing industries to assist in proofreading manuscripts. The report editor should also apply them to the design composition. When proofreading, it is best to remember:

1. All corrections should be made in the margins, if possible. The text should be marked to indicate the location of the correction. (Figure 1-12)
2. When there are three or more corrections on one line, both left- and right-hand margins should be used.
3. When correcting a letter, word, or series of words, mark through the existing letter, word, or words and write the substitute in the margin followed by /.
4. Notes to the typist, printer, or author should be written in the margin and marked "t," "p," or "a."

The following are the instructions a proofreader should use in this step:

Instruction	Mark
Delete	
Delete and close up	
Close up; delete space	
Move down	
Move up	
Move to the left	[
Move to the right	]
Equalize spacing between words	eq. #
Broken letter	X
New paragraph	¶
Run paragraphs together	no ¶
Let stand as typed/set	stet
Verify or supply information	(?)
Transpose letters or words	tr
Spell out abbrev.	(Sp)
Push space down	
Straighten type	=
Align type	‖
Run in material on same line	
Change (x/y) to shilling fraction	
Set as subscript	
Set as exponent	
Lower case Letter	lc
Capital letter	cap
Small capital letter	sc
Boldface type	bf
Italic type	ital

Roman type	rom
Wrong font	wf
Insert space	#
Turn letter	9
Period	⊙
Comma	^
Apostrophe	∨
Quotation marks	∨
Colon	:/
Semicolon	;^
Question mark	?/
Exclamation point	!/
Hyphen (-)	=/
En dash (–)	1/N
Em dash (—)	1/M
Two em dash (——)	2/M
Parentheses	(/)/
Brackets	[/]/

Polishing the design report is not a difficult task. It involves careful editing and rewriting to improve the quality of the written composition. Since both the report writer and the target audience may have different capabilities for understanding the report language, it is best to measure the material's overall *readability*. To do this, the author may apply any number of readability indexes to the passages. The applied indexes will then determine whether the readability is too simple or too complicated for the audience to understand.

AUTHOR:________________ PAGE NO.__________

123456789**1**123456789**2**123456789**3**123456789**4**123456789**5**1234

123456789**1**123456789**2**123456789**3**123456789**4**123456789**5**123456789**6**12345

1
2
3
4
5
6
7
8
9
10
11
12
13
14
15
16
17
18
19
20
21
22
23
24

Figure 1-12. This typical page may help in editing the final report. It is a 24-line (double-spaced) page with a 5½-inch-wide typing area.

For example, the management audience is comprised mostly of executives whose education is typically at the college junior or senior level, or higher. To satisfy this group, however, the readability level to aim for is a mid-high school range—slightly below their capability. Busy people, it seems, prefer to read material well below their capability because it takes less time for them to understand it. Such publications as *Time, Newsweek,* and The *Wall Street Journal* follow a similar concept.

The technical audience, on the other hand, may demand a higher readability level because of a usually higher education level. They seem to read, and "pick" at, material more thoroughly than others do, and a low readability level might insult them. The general audience is always a mixture. Reports written for them should be at the lowest readability level.

To assist in determining the readability of a design report, a report writer may wish to apply the *Gunning Fog Index* to selected paragraphs of the text. This readability index was developed by Robert Gunning and is one of the most applicable to the design professions. Select a writing sample of a hundred or more words and follow three basic steps:

1. Determine the average sentence length. Count the number of sentences in the passage. If there are two independent clauses in a single sentence, count each clause as a sentence. Then divide the total number of words in the passage by the number of sentences found in that passage. The result is the average sentence length.
2. Determine the percentage of difficult words. To do this, count the number of difficult words in the passage. Consider words having three or more syllables difficult. Do not count capitalized words; verbs made into three syllables by the addition of "-es" or "-ed"; or words that are a combination of short, easy words, such as *nevertheless* and *bookkeeper.* Difficult words that occur more than once should be counted each time they occur. Divide the total number of difficult words by the total number of words in the passage to obtain the percentage of difficult words.
3. Compute the fog index by adding the results obtained in Steps 1 and 2 and multiplying the sum by 0.4 (Figure 1-13)

This index corresponds to the grade levels of the U.S. School Systems. Grade 12 should be able to read at an index of 12, grade 10, at an index of 10, and so on. Always aim your index level slightly below that of the audience educational level.[2]

The final polishing may best be accomplished by asking a series of questions about the report. These questions are designed to identify problems before extensive resources are committed to production and include the following:[3]

Basic Organization

1. Are first things first?
2. Is each paragraph a complete, well-developed thought unit?

---

[2] Adapted from *The Technique of Clear Writing,* New York: McGraw-Hill Book Co., Revised Edition, ©1968 by Robert Gunning. Used with permission of the copyright owner.

[3] The checklist of questions is reprinted with permission from *Business Report Writing* by Phillip V. Lewis and William H. Baker, Grid Publishing, Inc., Columbus, Ohio, 1978, pp. 143-146.

LINCOLN'S GETTYSBURG ADDRESS

Fourscore and seven years ago our fathers brought forth on this continent a new nation, conceived in liberty and dedicated to the proposition that all men are created equal.

Now we are engaged in a great civil war, testing whether that nation or any nation so conceived and so dedicated, can long endure. We are met on a great battlefield of that war. We have come to dedicate a portion of that field, as a final resting-place for those who here gave their lives that a nation might live. It is altogether fitting and proper that we should do this.

But, in a larger sense, we cannot dedicate — we cannot consecrate — we cannot hallow — this ground. The brave men, living and dead, who struggled here, have consecrated it, far above our poor power to add or detract. The world will little note, nor long remember, what we say here, but it can never forget what they did here. It is for us the living, rather, to be dedicated here to the unfinished work which they who fought here have thus far so nobly advanced. It is rather for us to be here dedicated to the great task remaining before us — that from these honored dead we take increased devotion to that cause for which they gave the last full measure of devotion — that we here highly resolve that these dead shall not have died in vain — that this nation, under God, shall have a new birth of freedom — and that government of the people, by the people, and for the people, shall not perish from the earth.

Gunning Fog Index Exercise

Number of sentences	15	
Average length of sentences		12
Number of long words	22	
Percentage of total		8.6
Sum		20.6
Multiply by		0.4
Fog Index		8.24

According to this index technique, the Gettysburg Address delivered by President Lincoln was at a grade level of slightly more than eight. However, since the average person of that time only finished the fourth grade, he was speaking well over the heads of his audience.

Figure 1-13. This sample exercise determines the readability of a famous presidential speech. The Gunning Fog Index is a convenient method of determining the writing quality of the design report and its applicability to a target readership.

3. Are ample subheadings, or other display devises, used to quickly indicate the substances of the paragraphs or sections they begin?
4. Is each section of the report logical and well integrated?
5. Is there variety, originality, and force within the report?
6. Are facts organized to hold the reader's interest and to carry conviction?
7. Have all facts and figures been skillfully selected?
8. Has there been a comprehensive survey of all useful sources?
9. Has complete documentation been provided?
10. Is the development of the report consistent?
11. Is an outstanding weighting of the relative importance of the report's components evident?
12. Is each major consideration recognized and stressed?
13. Are the pros and cons presented in the report well balanced?
14. Is there a skillful relation of exhibits to the written text?
15. Do all arguments move ahead to a conclusion?
16. Is the relationship among pieces of evidence realistic?
17. Have all irrelevant details been eliminated?
18. Is there a clear synthesis of all data presented?
19. Are the main points of the report briefly summarized and displayed?
20. Are a summary and conclusion provided at the end of the report?

Writing Style

1. Is the structure of each sentence clear?
2. Do all words express the intended thought?
3. Is the language adapted to the vocabulary of the reader?
4. Is good syntax evident; that is, are word forms arranged to show mutual relationships in the sentence?
5. Are all sentences and paragraphs coherent; that is, do they stick together?
6. Are transitions provided for orderly procession of ideas?
7. Are the sentences generally short, with more few-syllabled words then many-syllabled words?
8. Have repetitions been eliminated?
9. Have you eliminated words that may be inferred from context or implication?
10. Is your report free from antagonistic words or phrases?
11. Is your report free from hackneyed or stilted phrases?
12. Is the tone of your report authoritative, courteous, and in tune with those receiving it?
13. Is the tone calculated to bring about the desired response?
14. Is the report easy to read and understand?
15. Have you been completely lucid?

Presentation Mechanics

1. Have you prepared a polished, clean final copy?
2. Does the physical appearance of each page create a favorable impression?
3. Have you left generous margins and given the reader an "easy eyeful of type"?
4. Have you used proper grammar and spelling?
5. Can the reader readily recognize the report's purpose?
6. Can you state a single purpose for each paragraph?
7. Can you state a single purpose for each sentence?
8. Can you state the purpose of each part of every sentence—word, clause, phrase?
9. Have you given the reader enough information all along the way?
10. Are the major divisions of the report (conclusions, recommendations, supporting data) easy to find?
11. Have you eliminated or relieved long explanations by a liberal use of illustrations and examples?
12. Are all charts, tables, and graphs clear and appropriate?
13. Is the report concrete rather than abstract?
14. Is a thorough proofreading evident? (Your report should be read, reread, edited, and reedited at least five times.)

Content and Analysis

1. Have you included all essential facts and provided sufficient detail?
2. Do you have adequate illustrative material to cover the points raised in the written text?
3. Have you made full use of supporting figures?
4. Are all facts directed to the reader's interest?

5. Is the accuracy of all factual information substantiated?
6. Are all statements in conformity with accepted rules and policies?
7. Has a proper synthesis of all data been made?
8. Is the report unbiased in its approach, equally for and against?
9. Are all arguments presented nonoverlapping?
10. Have you reached logical conclusions?
11. Have you considered all sensible alternatives?
12. Is your final summation of superior quality?
13. Are your recommendations unambiguous?
14. Have you reached a clear, definite decision that follows careful analysis?
15. Have all foreseeable consequences of recommended action been fully thought out, explored, discussed?

Miscellaneous

1. Did you know what was wanted and why when you started work on the report? (Always make sure you understand the overall situation.)
2. Have you dealt with a definite and limited problem?
3. Did you keep a clearly defined purpose in mind and stick to it?
4. Did you give serious thought to the need and temperament of the person for whom the report is prepared?
5. Has the right type of binding or fastening been used?
6. Does the title express the nature and value of the report and make the receiver want to read it?
7. Have you provided a table of contents or an index so the reader may quickly locate desired material?
8. Is the report within an acceptable or a required lengthy?
9. Have enough copies been made for distribution?

2. Structure

The design report is an instrument of organization and a means of relating technical information to people. Because it is so important to successful design and planning operations, it should not be treated as simply a cosmetic addition. Poorly written and produced reports are very often associated with poor-quality design. Clients feel that if a consulting team puts so little effort into a finished report, they might employ the same level of effort in the development program. The report document is the last, and often strongest, link in the chain of communication. Treat it carefully, but produce it effectively.

One basic characteristic governs the structure of all design reports—they are written to be read. All of the documents prepared by a consulting organization convey some kind of information to someone in order to attain some type of objective. Each document, therefore, should have a specific direction, a defined content, and an organized format.

The *direction* of the design report can be either vertical or horizontal. *Vertical reports* are directed most often at designers within the same organization or designers of different organizations. They are usually more technical in nature, less descriptive in written documentation, and are subject to interpretations by trained professionals.

Horizontal reports, on the other hand, are just the opposite. They are usually written for a client and are the most important in terms of the required production commitment. This type of material must be less technical in nature, and more descriptive in narrative and graphics for comprehension by nontechnical readers; it must also have a specific interpretation as an overall goal. (Figure 2-1)

The *content* of the report must be geared to (1) the presentation of facts, (2) a review of research techniques or design processes, and (3) the outlining of specific recommendations for design actions. Each is an important ingredient of structure and will guide the final preparation of the composition.

The *format* of the design report is usually the "image" factor generated by the production team. Unfortunately, first impressions are a very critical issue within the design industry, so the format may affect the acceptance or rejection of the information. A bad image will often kill a good design, while a fair design may be accepted if the graphics are more easily understood by the target readership.

Since it is often difficult for the public at large to compare the quality of competing designs, the report may be the only available medium for comparison. One set of drawings looks much like the other during a proposal competition, and many good ideas have often gone astray or have been rejected because the information was misunderstood or misdirected. Problems arise when the designer of the report fails to "package" the information in such a way that the nontechnical reader or listener can properly comprehend what is involved. To effectively communicate the ideas of the consulting

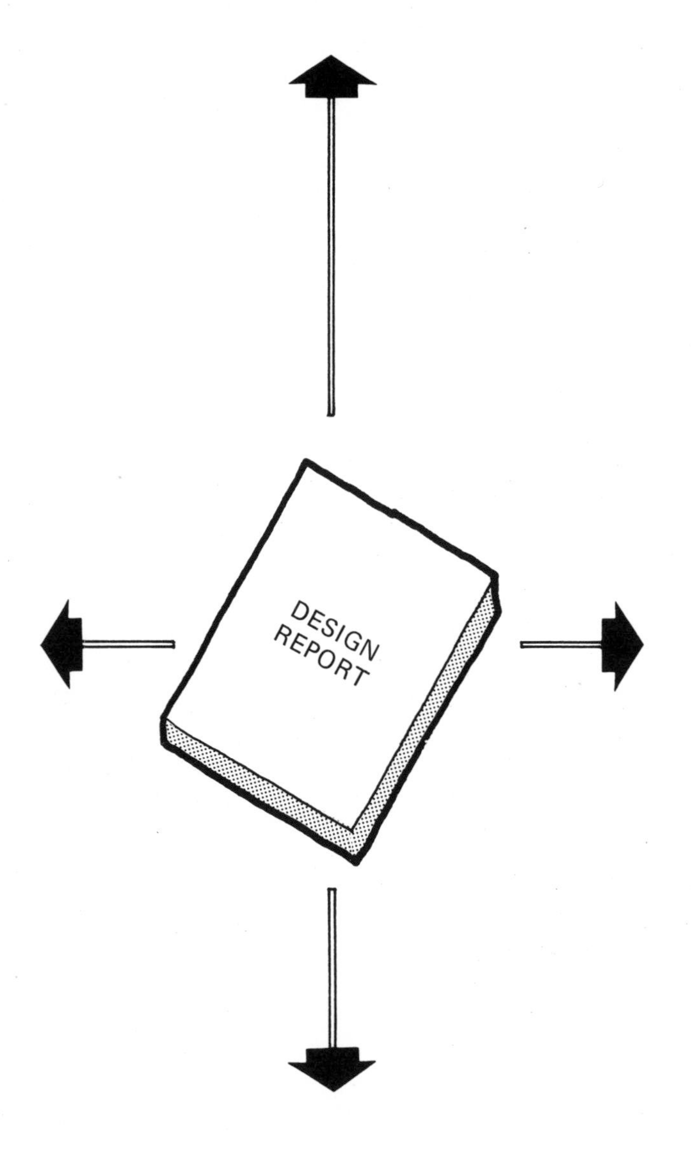

Figure 2-1. The direction of the report will determine its basic readability. The vertical report will be the most technical because it must communicate detailed information to other designers. Horizontal reports, on the other hand, are written for the client and are less technical.

team, the report should have the following essential characteristics:

1. *It must attract attention to the important issues of the design program.* Each important ingredient of the design program must be fully explained to the client in nontechnical language. This will enhance the comprehension of the information and allow the client to feel the goals and objectives are attainable.
2. *It must establish the identity of the project.* A successful design campaign is always improved with effective public relations. Through the strategic placement of illustrative graphics and supportive text, the design report can supply a project with identity and a sense of "belonging" to a client or community group.
3. *It must develop and hold interest in the project.* Too often, clients leave a presentation with a sense of confusion about the objectives of a design program. A carefully written and produced report should eliminate this confusion and maintain adequate resource commitments until the project is realized.
4. *It must generate the actions necessary for the completion of the project.* After reading the design report, the client should be motivated to undertake and complete the work outlined by the designers.

THE BASIC COMPONENTS

A well-produced design report is comprised of several important parts that should be combined into an effective communication package. Each part should act independently of the other parts, but when blended together, should create a desired impact upon the target audience. The three main units of the report are, the front matter, the main body, and the back matter.

The *front matter* includes the following:

- Cover (front)
- Title page
- Preface or foreword
- Acknowledgments
- Contents
- List of illustrations

The *main body* includes the following:

- Introduction
- Chapter components
- Discussion or summary
- Recommendation

The *back matter* is comprised of the following:

- Appendix
- Glossary
- List of references
- Bibliography
- Index
- Cover (back)

The Front Matter

Front Cover

The readership of a design report is somewhat captive, usually because it includes a client or lay group with a specific interest in the material contained within the document. Nevertheless, the front cover of the report remains as the most important page. It must attract attention and create interest in the contents that follow. Its main job is to be pleasing, to advertise the contents without contradiction, and to be in character with the design program. (Figures 2-2 and 2-3)

Differences between good and bad cover design are usually very difficult to determine. An artist may spend hours on the composition and never reach the desired objective. The value judgments of those who will review the finished work are many and will vary with each individual. However, successful cover designs are usually simple in composition, while unsuccessful ones are often complicated and require more "reading" than is necessary.

The front cover should contain only the title of the work or program. The name of the organization, designer, or authors of the document should be placed elsewhere. Information overkill only increases the complexity of the product.

The title should be simple, easy to read, and follow the natural eye-flow of the average reader. Confusing and obscure letter arrangements create negative reactions from the readers, and the interest in the contents may be seriously diminished. Creativity is important, but caution is the rule. (See section on *Designing with Type* for the psychological aspects of eye-flow) (Figure 2-4)

If artwork is to be used it should be simple, creative, and, most of all, nonconflicting with the main title. Variations might include:

1. A full-blend photograph with title copy surprinted or dropped-out in the picture area. (Figure 2-5)
2. A cover with an illustration as part of the title copy. (Figure 2-6)
3. Several areas of illustration or photographs in association with the main copy. (Figure 2-7)

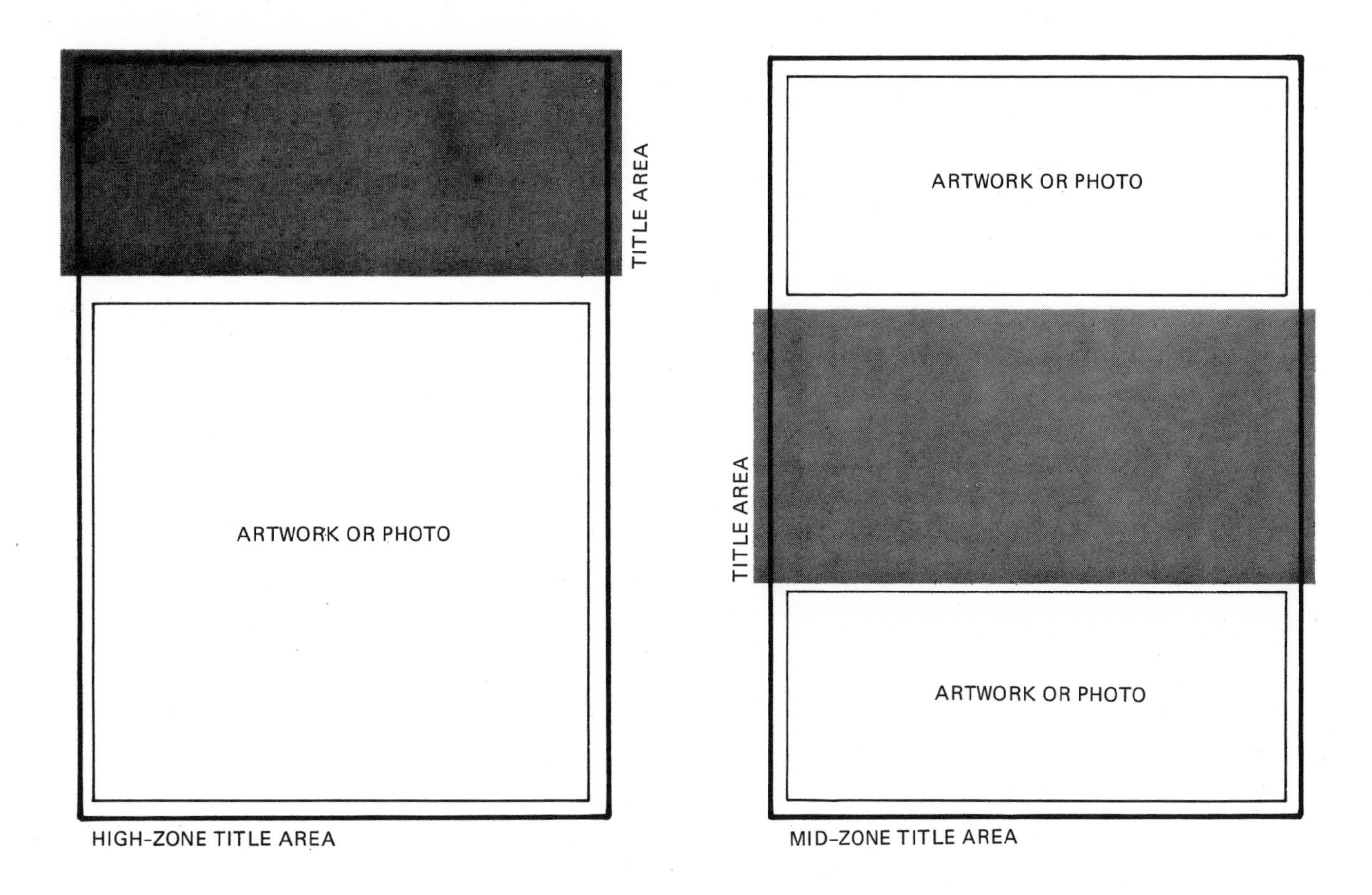

Figure 2-2. The *high-zone title* is similar to the format used by most magazines. The artwork area is used to accent or support the content theme. The *mid-zone title* is often used for design reports. The artwork areas divde the graphic support of the content theme and are best left unfilled to prevent confusion.

Figure 2-3. The *base-zone title* is the reverse of the magazine format. It offers a large graphics area for artwork or photographs and provides a "foundation" appearance for the title.

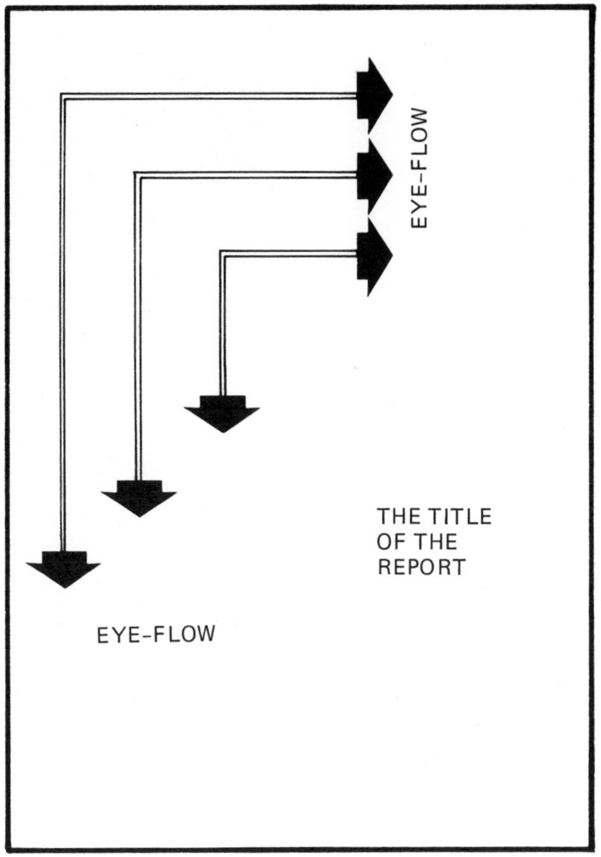

EYE-FLOW

THE TITLE
OF THE
REPORT

EYE-FLOW

Figure 2-4. Eye-flow is important for the visual decoding of the type styles used in printing. Be sure the right-hand side of the type is more readable than the left, and the upper part more readable than the lower.

Figure 2-5. Courtesy, The University of Nebraska.

UNITED STATES DEPARTMENT OF THE INTERIOR
Heritage Conservation and Recreation Service

Figure 2-6. Courtesy, U.S. Department of the Interior.

Figure 2-7. Courtesy, Theraplan, Incorporated.

Figures 2-8 through 2-30 show sample covers that have been developed for design reports.

Title Pages
The title page of the design report provides the detail information essential for the identification of both the document producers and recipients. The location of the design organization, the date of the report, the author(s), and the main title should also be included on this page. Figures 2-31 through 2-33 are samples of title pages.

Preface or Foreword
A short statement of philosophy on the material in the report document is always helpful to the reader. It helps to set the scene or mood for the material that will follow. A comment on how the content will assist the reader is also helpful.

Acknowledgments
Credit should be given to persons or organizations that may have assisted in writing or producing the final document. Advice that was given, any materials that were provided, or permissions granted for reprinting should be listed in this section. This is not a legal requirement, but it is a major professional courtesy.

Table of Contents
This unit identifies the major portions of the text that are used to present the design program. Each chapter, or part, heading listed here should be consistent with the heading used in the text area. Subheadings may be used if they are not too complicated and if they do not conflict with the main headings. (Figure 2-34 through 2-38)

List of Illustrations
Each of the supportive illustrations that is used in a design report should be listed or credited separately on this page. The title of the illustration, author, source, and page number should be presented. Tables, charts, and figures should also appear in this manner.

The Main Body

Introduction
This is the first of the major parts of the design report and its purpose is multidimensional. First, it should capture the reader's attention and provide background information on the subject material. If the report is about a major state park facility, for instance, the reader should be informed about its location, existing facilities, benefits, and relationship to other nearby units before the main body of the report occurs.

Second, it should state the purpose of the report in terms of benefits to the client, organization, or citizen's group. Explaining why a report is needed may often eliminate objections to the development program proposed in later sections. Third, a short history of the project or design problem, and how the consulting team approached the solutions, is always helpful. Any assumptions the design team might have regarding the solutions outlined in the final document should appear in the introduction. (Figures 2-39 through 2-41.)

Figure 2-8. Courtesy, Johnson, Johnson and Roy, Inc.

Figure 2-9. Courtesy, Myrick, Newman and Dahlberg and Partners, Inc.

wilderness recommendation

Figure 2-10. Courtesy, U.S. Department of the Interior.

United States Department of the Interior
Heritage Conservation and Recreation Service

SEPTEMBER 1978

Figure 2-11. Courtesy, U.S. Department of the Interior.

Figure 2-12. Courtesy, Myrick, Newman and Dahlberg and Partners, Inc.

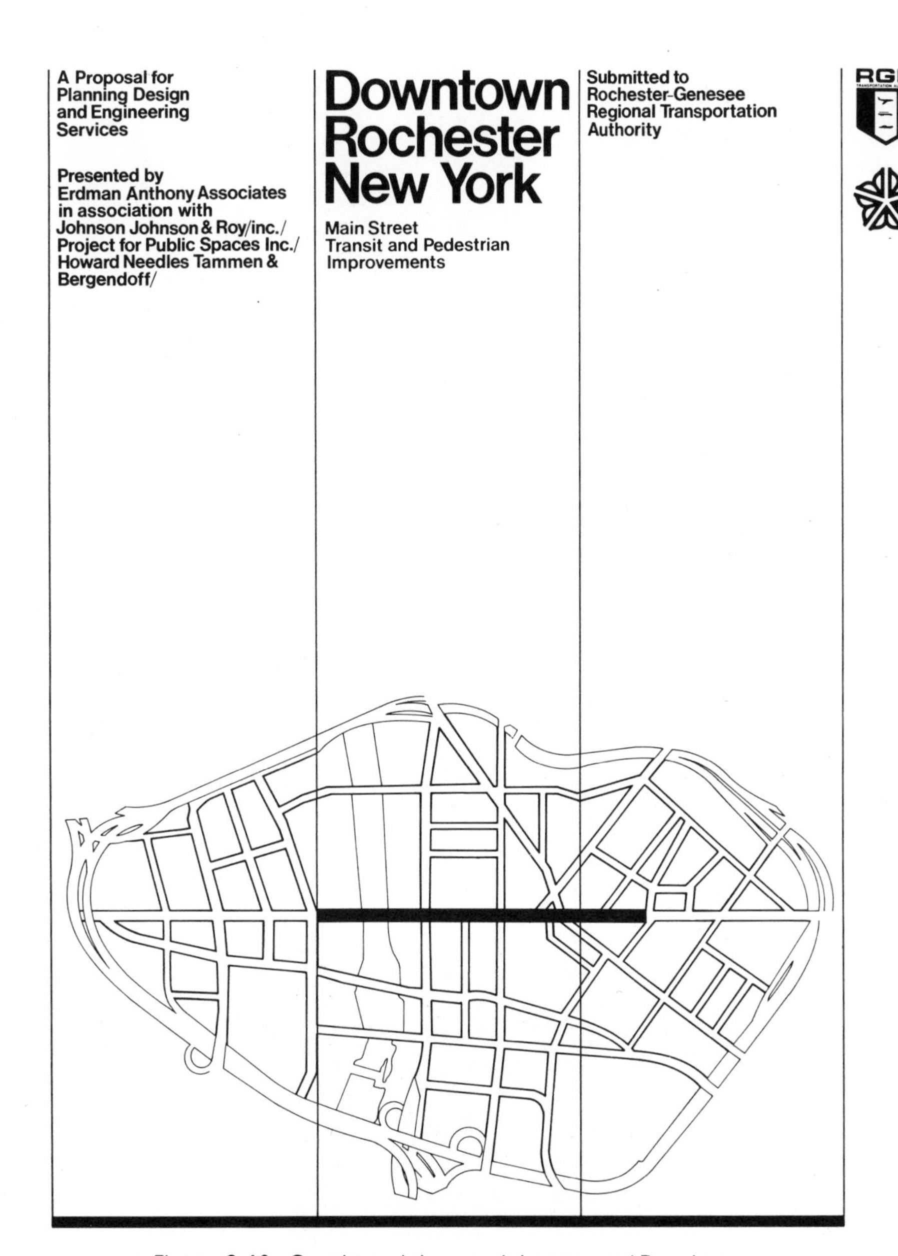

Figure 2-13. Courtesy, Johnson, Johnson and Roy, Inc.

people
are asking
about...

displaying
the symbol
of access

Figure 2-14. Courtesy, The President's Committee on Employment of the Handicapped.

MONOGRAPHS

Figure 2-15. Courtesy, The University of Nebraska.

COUNCIL GROVE: HISTORIC CONSERVATION

Figure 2-16. Courtesy, Kansas State University.

DOE/CA/10879-01

CONSUMER ENERGY ATLAS

June 1980

Prepared for:
U.S. Department of Energy
Office of Consumer Affairs
Under Contract No. AC03-80/R/0879

Figure 2-17. Courtesy, U.S. Department of Energy.

The Atchafalaya
America's Greatest River Swamp

Figure 2-18. Courtesy, U.S. Fish and Wildlife Service.

Figure 2-19. Courtesy, Johnson, Johnson and Roy, Inc.

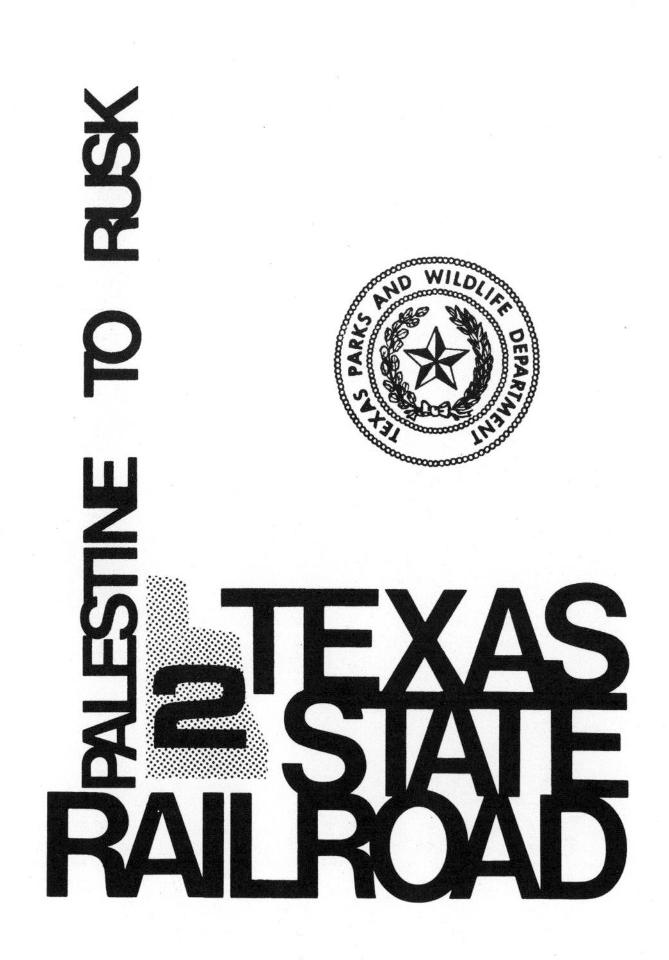

Figure 2-20. Courtesy, Texas Parks and Wildlife Department.

Figure 2-21. Courtesy, The University of Nebraska.

Figure 2-22. Courtesy, The University of Nebraska.

Figure 2-23. U.S. Department of the Interior.

Figure 2-24. Courtesy, Texas Parks and Wildlife Department.

Figure 2-25. Courtesy, The University of Nebraska.

SCROUNGING

UNITED STATES DEPARTMENT OF THE INTERIOR
Heritage Conservation and Recreation Service
April 1980

Figure 2-26. Courtesy, U.S. Department of the Interior.

Figure 2-27. Courtesy, The University of Nebraska.

Figure 2-28. Courtesy, The University of Nebraska.

VOLUNTEER HANDBOOK

UNITED STATES DEPARTMENT OF THE INTERIOR
HERITAGE CONSERVATION AND RECREATION SERVICE
SEPTEMBER 1978

Figure 2-29. Courtesy, U.S. Department of the Interior.

Government Periodicals and Subscription Services

Superintendent of Documents
U.S. Government Printing Office
Washington, D.C. 20402

Price List 36
Edition 173
May 1979

Figure 2-30. Courtesy, Superintendent of Documents, U.S. Government Printing Office.

WILDERNESS RECOMMENDATION

Badlands
National Monument
South Dakota

August 1972

United States Department of the Interior / National Park Service

Figure 2-31. Courtesy, U.S. Department of the Interior.

PATHWAYS & PADDLEWAYS

A
Trails & Scenic Waterways
Feasibility Study

Conducted by:

Wayne D. Oliver
Clyde M. Biggs
Candy Abshier Finney
Ron Thuma

Under Direction of:

Ron D. Jones, Director of Planning

Published by:

Texas Parks and Wildlife Department, Austin, Texas
August, 1971

Figure 2-32. Courtesy, Texas Parks and Wildlife Department.

VETERANS ADMINISTRATION NATIONAL CEMETERY

Fort Custer, Michigan

Prepared by: Johnson Johnson & Roy/inc.   303 North Main Street Ann Arbor Michigan 48104

Figure 2-33. Courtesy, Johnson, Johnson and Roy, Inc.

contents

1 introduction

THE PURPOSE OF THE PROJECT 2
PROJECT ORIENTATION 4
ITS HISTORY AND SETTING 7
VISITATION 8
GOALS AND OBJECTIVES 8

2 the interpretive plan

THE PUBLIC USE AREAS 11

Woodridge
Bloomington
Rockhaven
Southcove

THE OLD BLOOMINGTON TOWNSITE 15

Townsite Courtyard
Townsite Museum

THE INFORMATION CENTER 20

The Exhibit Area
The Interpretive Trails

INTERPRETIVE SIGNAGE 47

3 implementation

PHASING 49
COSTS 50
OPERATION AND MAINTENANCE 50

4 appendix

DATA/MATERIALS 54
THE PLANNING TEAM 55

Figure 2-34. Courtesy, Theraplan, Incorporated.

Contents

Introduction	3
Part 1 Background	7
Location and History	8
Environment	10
Part 2 Master Plan	15
Concept	16
The Visitor Experience	17
Features	18
The Entrance	20
Administration	22
Committal Service Shelters	24
Service-Maintenance Facility	26
Part 3 Development Guidelines	29
Landscape Character	30
Design Criteria	30
Treatment of Edges	31
Site Guidelines	32
Land Management	34
Summary	35

5

Figure 2-35. Courtesy, Johnson, Johnson and Roy, Inc.

Contents

INTRODUCTION 1

STRATEGY PLANNING 3

- Defining Goals and Objectives 3
- Organizing 4
- Inventorying Needs and Resources 6
- Analyzing the Resources 7

IMPLEMENTATION 10

- The Approach 10
- Outright Donations 11
- Events-Sales-Services 16
- Public Relations Activities 18

CASE STUDIES 20

- Wild Basin Wilderness (Christmas in June) 20
- Otis Park, Colorado Springs, Colorado (Special Park) 22
- The Platte River Greenway (A $6 Million Effort) 23
- Alamogordo Park and Recreation Department (Private Sector Involvement) 24
- Cyclone Area Community Center, Iowa (Dedication, plus) 25
- Columbine Area Playground (The Four-hour Fundraising Campaign) 26
- East Oakland Youth Development Center (Corporate Initiative) 27

SOURCES OF INFORMATION 29

The examples on the cover are documented in the case studies beginning on page 20. Read "Strategy Planning" and "implementation" to find out how you can easily do the same thing.

Figure 2-36. Courtesy, U.S. Department of the Interior.

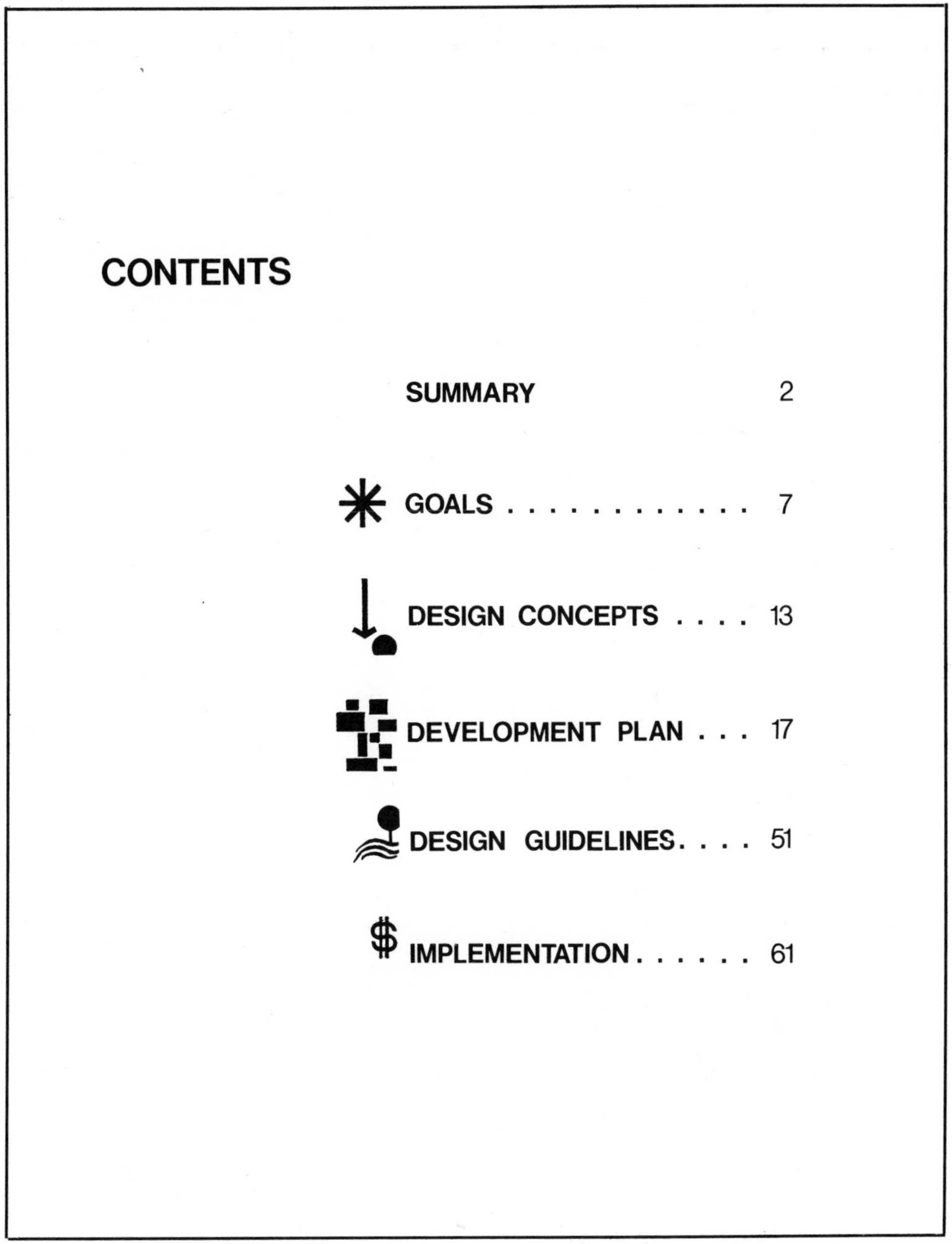

CONTENTS

SUMMARY 2

GOALS 7

DESIGN CONCEPTS 13

DEVELOPMENT PLAN . . . 17

DESIGN GUIDELINES. . . . 51

IMPLEMENTATION 61

Figure 2-37. Courtesy, Myrick, Newman, Dahlberg and Partners.

TABLE OF CONTENTS

Introduction 1

Past 6

Present 11

Surveys 22

Districts 24

Downtown 31

Conclusions 44

Appendix 50

Figure 2.38. Courtesy, Kansas State University.

Introduction

The rural landscape of Southern Michigan embraces a pleasant mosaic of vegetation, water features, and land forms. The discerning observer will recognize this landscape as the product of both nature and man, reflecting a complex history dating from the last glaciation and spanning Indian occupation, the pioneer lumbering era, the agricultural economy and, finally, the birth and growth of American industry.

This area is the setting for the Fort Custer National Cemetery, one of more than 100 national cemeteries managed by the Department of Memorial Affairs of the Veterans Administration. The cemetery provides burial benefits for veterans as established by the 37th Congress in 1862 and the National Cemeteries Act of 1973. The Master Plan provides for sufficient gravesites to last far into the next century.

Visitors will notice that this facility differs intrinsically from the vast formal character of other national cemeteries. Here, care has been taken to conserve natural areas and to fit the cemetery into the site's many private clearings and meadows.

This brochure describes the environment of the site and its history, the Master Plan for the development of this national cemetery, and guidelines for implementation. It outlines the transformation of a rural site, familiar to many Midwest families as a former military training center, to a national cemetery, a pastoral final resting place for Midwest veterans.

3

Figure 2-39. Courtesy, Johnson, Johnson and Roy, Inc.

Introduction

Fundraising is a selling plan. It depends on communicating purposefully with people by asking them for support. There are many methods of fundraising, many rules, and just as many exceptions to the rules. Whether your fundraising effort is to last a year or a month, it involves some basic lessons. This booklet discusses the basic "how to's" of fundraising and provides case examples of ideas, alternatives, and methods that have proved successful.

Before getting into the "how to's," it is important to understand the purpose of fundraising and why it is beneficial to expend the time and energy in organizing a fundraising program. Whether you represent a public institution such as a city park and recreation department, a small neighborhood group, or a national nonprofit organization, you are only part of the larger whole. One purpose of fundraising is to effectively coordinate efforts of the government, private, and community sectors to benefit the whole.

There are many methods of fundraising.

Many agencies and organizations face the problem of operating on a limited and inadequate budget. In the field of parks and recreation, for instance, all encompassing changes in values, technology, energy use, and the economy are challenging the traditional approach to providing open space and leisure services. The May 1979 issue of *Parks and Recreation* says that park and recreation departments "must now compete as never before for the budget." The reality that cuts into municipal budgets is drastically limiting what government can provide, encourages a "new spirit of self help, community involvement, and consciousness [that] is emerging in many places. The spirit recognizes the limits of government in

Figure 2-40. Courtesy, U.S. Department of the Interior.

INTRODUCTION

WHAT IS HISTORIC CONSERVATION?

Historic conservation in its most dynamic form is much more than sentimental preservation of recognized historic landmarks. It involves the successful blending of the old with the new, and can be the source of community pride. Utilization of historic resources is limited only by the imagination.

Let's look more closely at the exact differences between historic *conservation* and historic *preservation*. Conservation is the policy of *adaptation of new functional uses* so that historic landmarks may fulfill an active role in modern times. Preservation is the *setting aside* of important historical landmarks, so that they may be enjoyed by all. Examples of historic preservation are the landmarks designated along the Historical Trail in Council Grove. These landmarks have been maintained in their original form. Aside from being observed, however, they are no longer functional.

Many times it is not economically feasible to maintain historical landmarks in this manner. Consider historical buildings, for example. In our cities everywhere, many of our century-old buildings are being torn down on a regular basis. Of course it is not possible to set aside each old building as a museum of some type. Nor is it feasible to save a 3-story building built in the 1800's, when a 20-story building utilizing more of the same space will provide so many more offices. As a result, products of our cultural heritage are being replaced by steel and glass skyscrapers. Many citizens are more than concerned about this loss. Too often this concern is dismissed in recognition of "progress."

Certainly there is no sure-fire solution to the problem. It is true that the skyscraper is very much a part of our modern American culture. However, the buildings of our past were the predecessors to the skyscrapers, and the two are not entirely incompatible.

Many times, a compromise may be reached between the preservation of an old building and the demands met by a new building. If one or a group of old buildings can functionally meet the demands of our society today, they may be saved through the adoption of a conservation policy. When these older structures become economically viable and functional, the result is a successful blending of the old with the new. An example of this policy is the River Quay area of Kansas City, where old warehouses have successfully been converted to shops, restaurants, clubs, etc.

An area in Council Grove where a historic conservation policy might be applied is the downtown section of buildings built in the late 1800's. Perhaps this area is the most overlooked "landmark" in the community. It is important to remember that a landmark does not necessarily have to be set aside. Many times it can serve as a functional part of the community.

The basis of a unified conservation policy is the survey and analysis of the many elements of the community. This includes not only buildings, but building arrangement, open spaces, street layout, vehicular and pedestrian circulation, and even small details that often go unnoticed. The fabric resulting from these interwoven elements can be described as the overall "character" of the community. A conservation policy must keep in close harmony with this "character."

In the following pages are the findings from our group's survey of Council Grove. From this survey, we determined guidelines for the adoption of a historic conservation policy applicable to Council Grove.

2

Figure 2-41. Courtesy, Kansas State University.

Chapter Components
This is the main portion of the design report document. It contains information on the specific topics of the design or development programs and should be treated with special emphasis. Each division within the section, such as design methodology, analysis results, or implementation strategies, will appear more professional if each has a short introductory statement. Follow this treatment with the text of each topic, and conclude with a short summary. If the narrative is lengthy and there are a number of divisions, a transitional statement from one division to the next is always helpful.

Discussion or Summary
This material should present each of the major facts of the report in a clear and concise manner. Failure to mention each important fact and its possible impact upon the outcome of the program may result in a misunderstanding of the project. A short lead-in statement and an itemized list are helpful to most readers.

Recommendations
This section is usually a very important part of any report effort. Statements made here will determine much of the success or failure of the presented program. If a recommendation is made, it should be supported with both the positive and negative aspects of the issues in the document.

The Back Matter

Appendix
This part of the report is comprised of information that both supports the major statements made in the main body of the report and provides data that will expand the reader's understanding. Items that should be included in the appendix are: (1) a sample questionnaire of the design survey; (2) the tabulated results of the survey; (3) letters from individuals or organizations that may support the intended program; or (4) charts, tables, and graphs that assist in clarifying information for the reader.

Glossary
This area of the report is often overlooked by many designers. Correct terminology is important for accurate interpretation of report information, and this section can summarize the technical terms more effectively for increased reader comprehension. What may be a common term to a professional may be something else entirely to management or a private citizen. The glossary is an excellent tool for bridging this data gap.

List of References
There are legal restrictions on the use of materials obtained from outside sources. To prevent a violation of these restrictions, a list of sources is a necessity. Drawings, photographs, tables, or other pertinent outside material should be credited in this list.

Bibliography
This section should list the general sources of information used to develop the design report. Personal interviews, films, handbooks, letters, or unpublished technical works by colleagues should be listed in a bibliography.

Index
If the report is exceptionally long, it is best to have an index to facilitate the location of important information. It is not necessary to list every term or heading in this component, but conveni-

ent access to key issues or facts will be a valuable aid to the reader.

Back Cover
Most design report back covers are blank, which unfortunately creates waste and extra costs. The back cover can be used for listing the company name, logo, or names of design team members in such a manner that the entire composition maintains continuity. The samples in Figures 2-42 through 2-44 illustrate this important point.

PAGE COMPOSITION

Organizing the composition of the report page is probably one of the most difficult tasks a designer will undertake. Choosing a margin, selecting the photographs or line art, and placing them together where they will "read well" involve more steps than one might imagine. One of the best ways to accomplish this is to develop a graphic zone, or *grid*, at the beginning of the composition process. The grid will be of assistance in the paste-up operations for individual page layouts. It should not, however, be used as a straitjacket, but more as a creative aid to maintaining a sense of balance and continuity throughout the report.

With a grid base and a light table, the report designer can maintain the discipline that will be required to provide the flow from page to page of the report. Grids allow for a better understanding of complex information. Concept sketches, tables, construction schedules, financial material, and technical data can be arranged more easily on a page with a grid. Figure 2-45 shows a standard arrangement for a report page grid.

The layout of the individual page of a report is as important to the overall document as is the composition of a detailed construction drawing to the building of structures and sites. Each must be consistent with the communication needs of the program and each must be easily understood. Careful consideration, therefore, should be given to the composition of the page columns, which in turn will determine the most suitable margins for the copy. Figures 2-46 through 2-56 show typical column arrangements for a report document.

Page Sizes and Margins

A commonly used page size for the design report is the 8½ × 11 upright unit. It offers the most economical approach for the production of design information and can be altered to create numerous variations for composition. Margins are arbitrary and their selection should be based upon eye-scan and the needs of the report, but will include one or all of the following:

1. *Flush left.* The left-hand margin is aligned. This is the best choice for natural word flow and can be set or prepared on a standard typewriter. The right-hand portion is ragged. (Figure 2-57)
2. *Justified.* This is flush on both sides of the text and is often used for books and magazines. Corrections are often difficult because of the exact number of spaces needed to fill the line width. (Figure 2-58)
3. *Flush right.* The right-hand margin is aligned. This margin is best for short descriptions, such as art, photo captions, or titles. It may, however, cause confusion and problems with eye-scan. (Figure 2-59)

Figure 2-42. Courtesy, U.S. Fish and Wildlife Service.

Figure 2-43. Courtesy, Texas Parks and Wildlife Department.

Figure 2-44. Courtesy, The University of Nebraska.

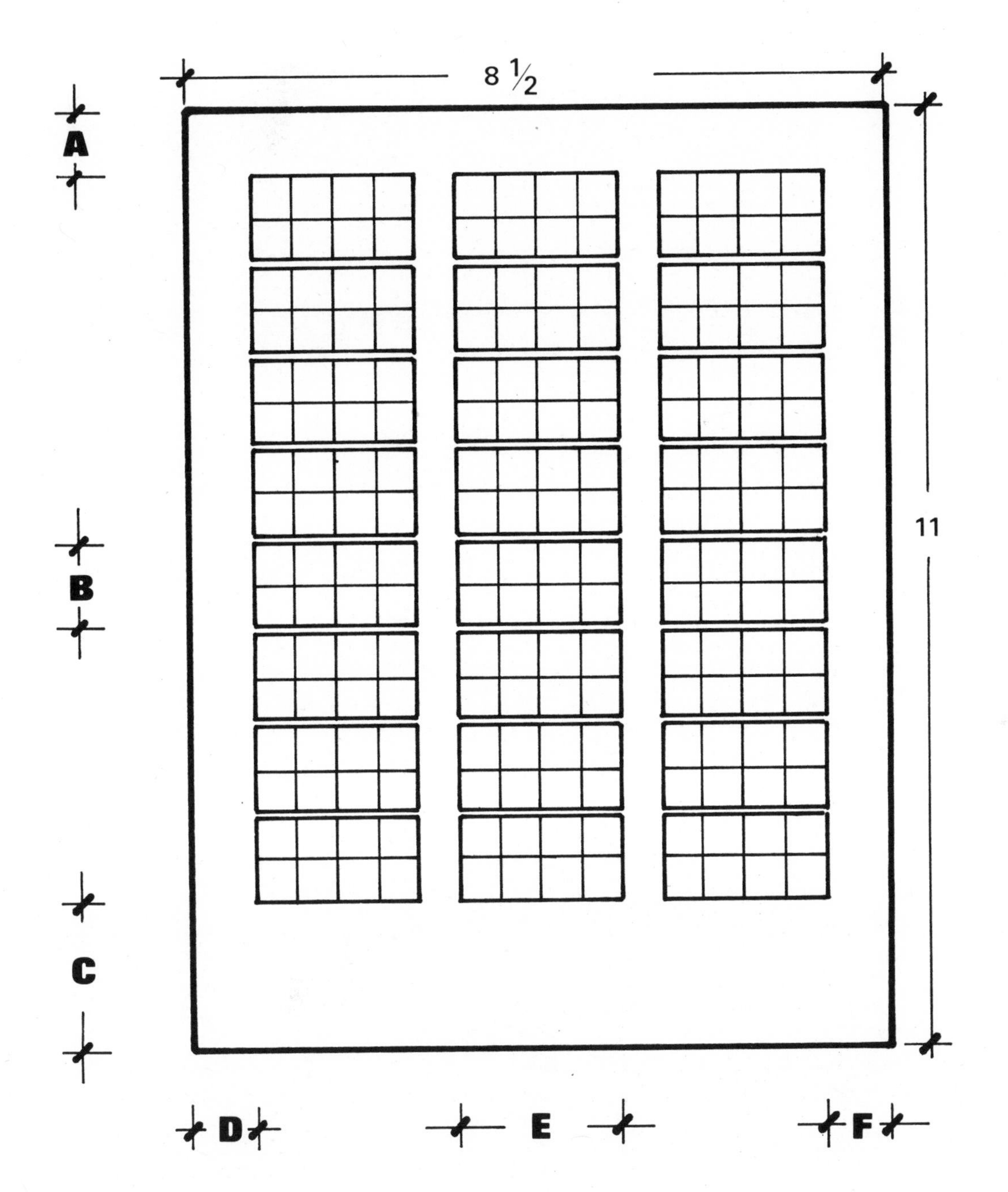

Figure 2-45. A = Margin, B = Grid unit, C = Base margin, D = Margin, E = Grid column, and F = Margin.

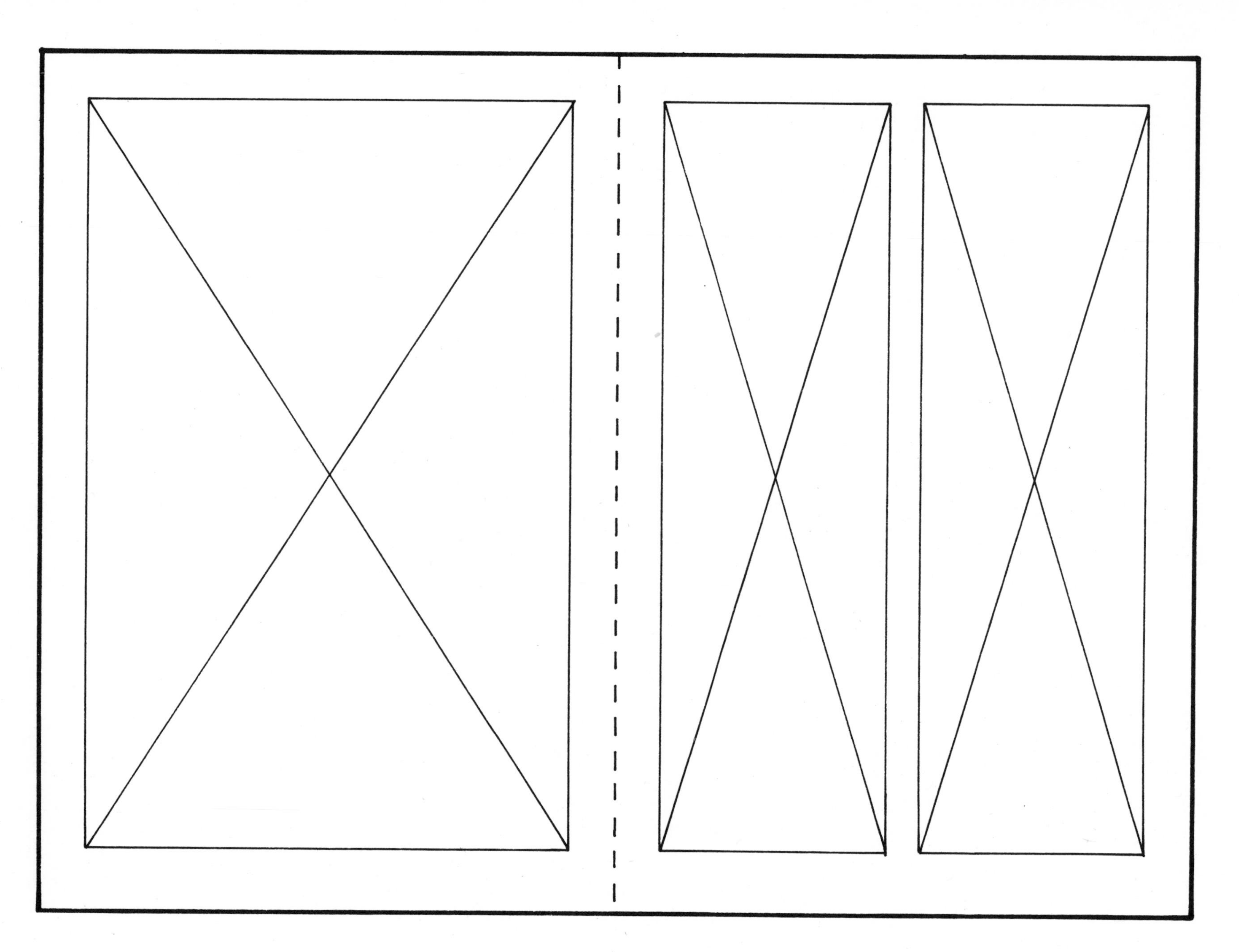

Figure 2-46.

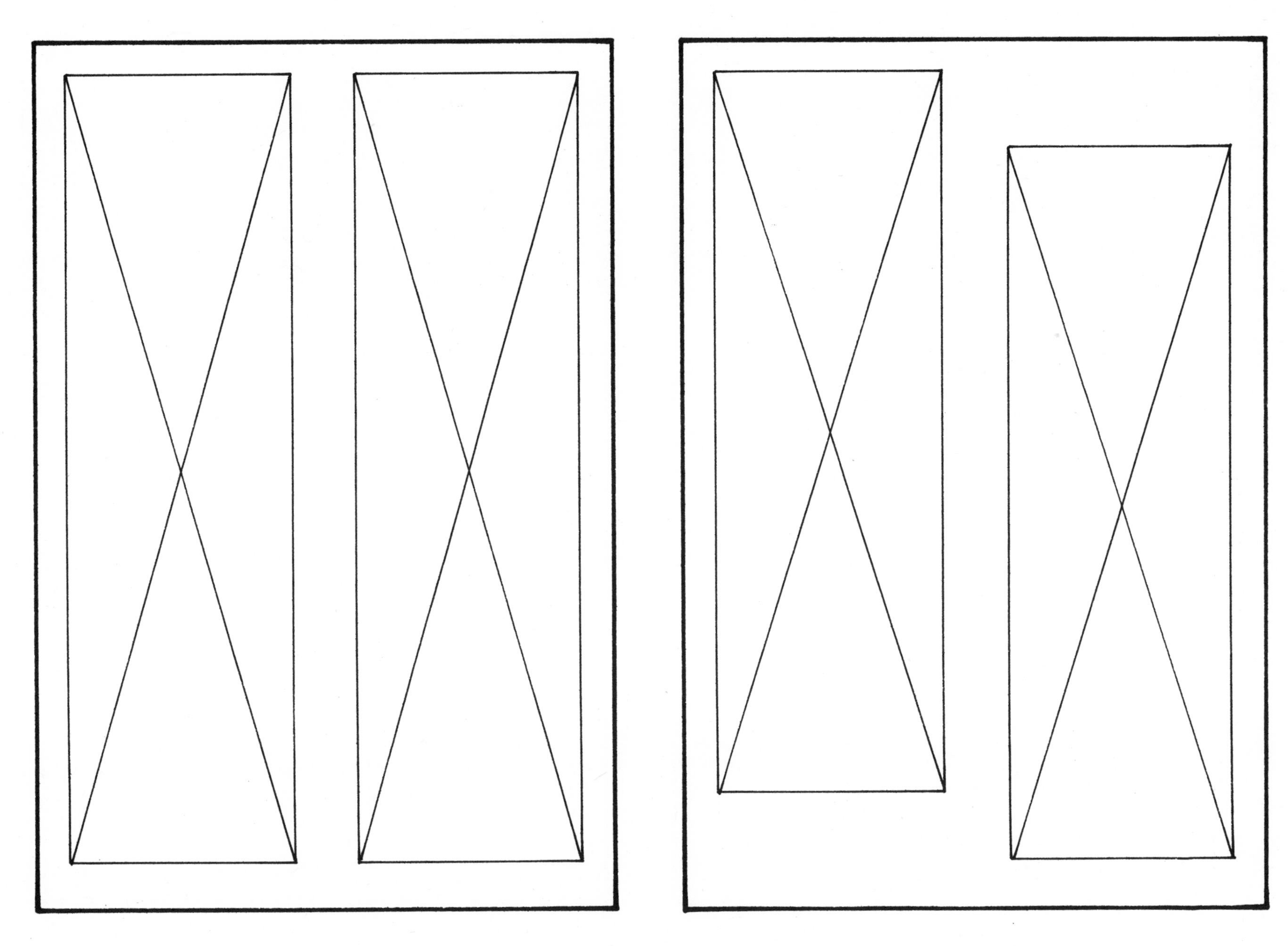

Figure 2-47.

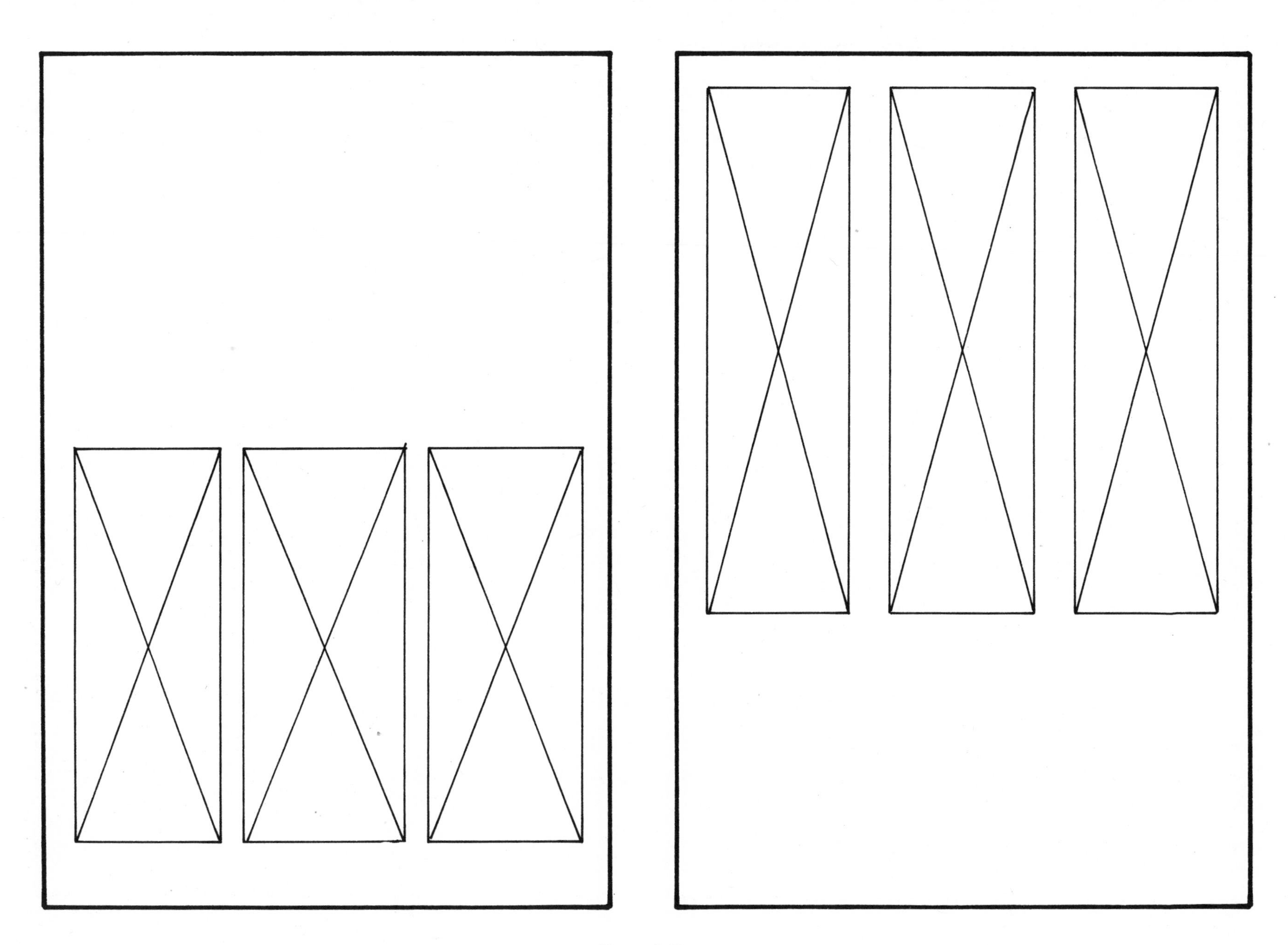

Figure 2-48.

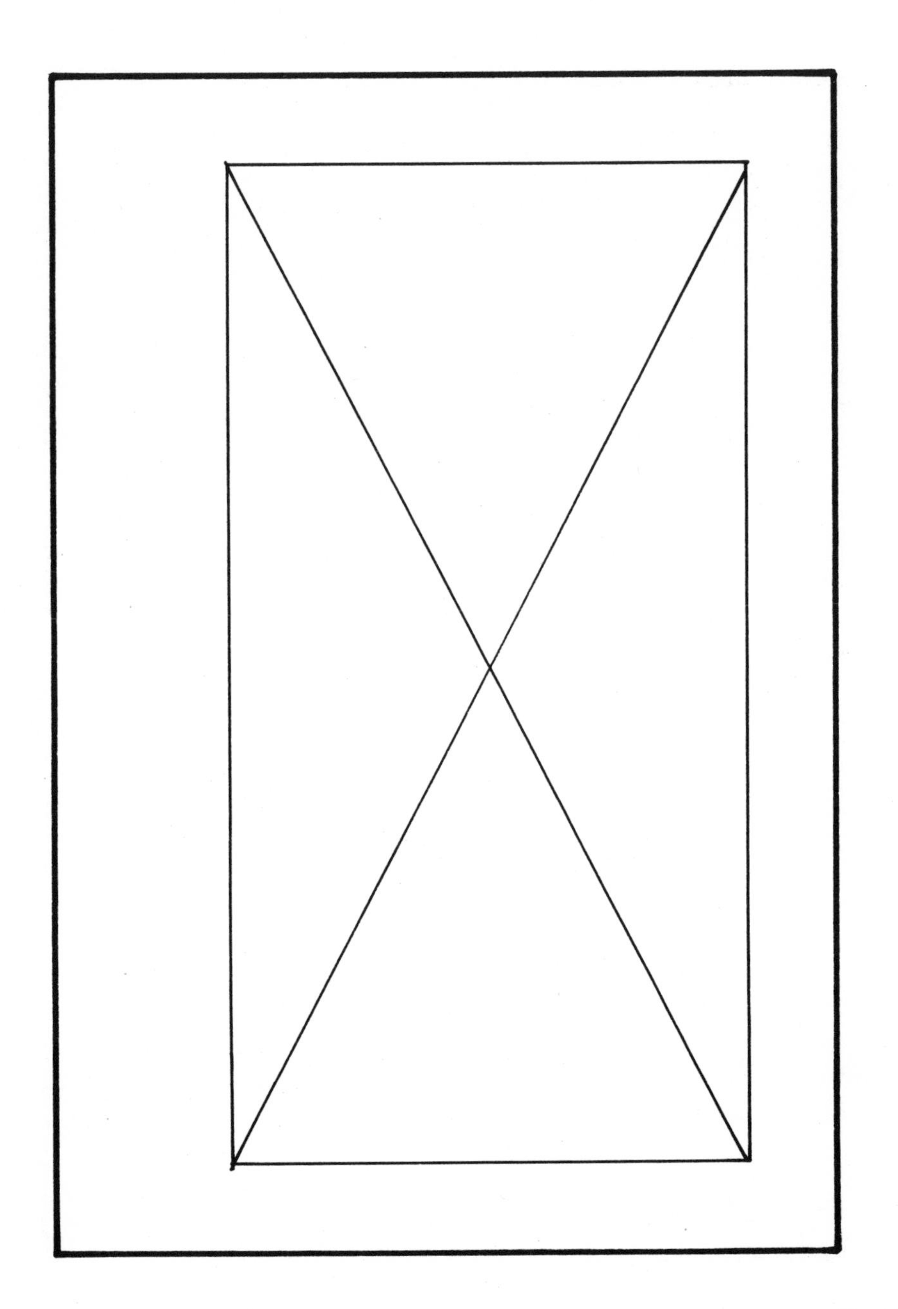
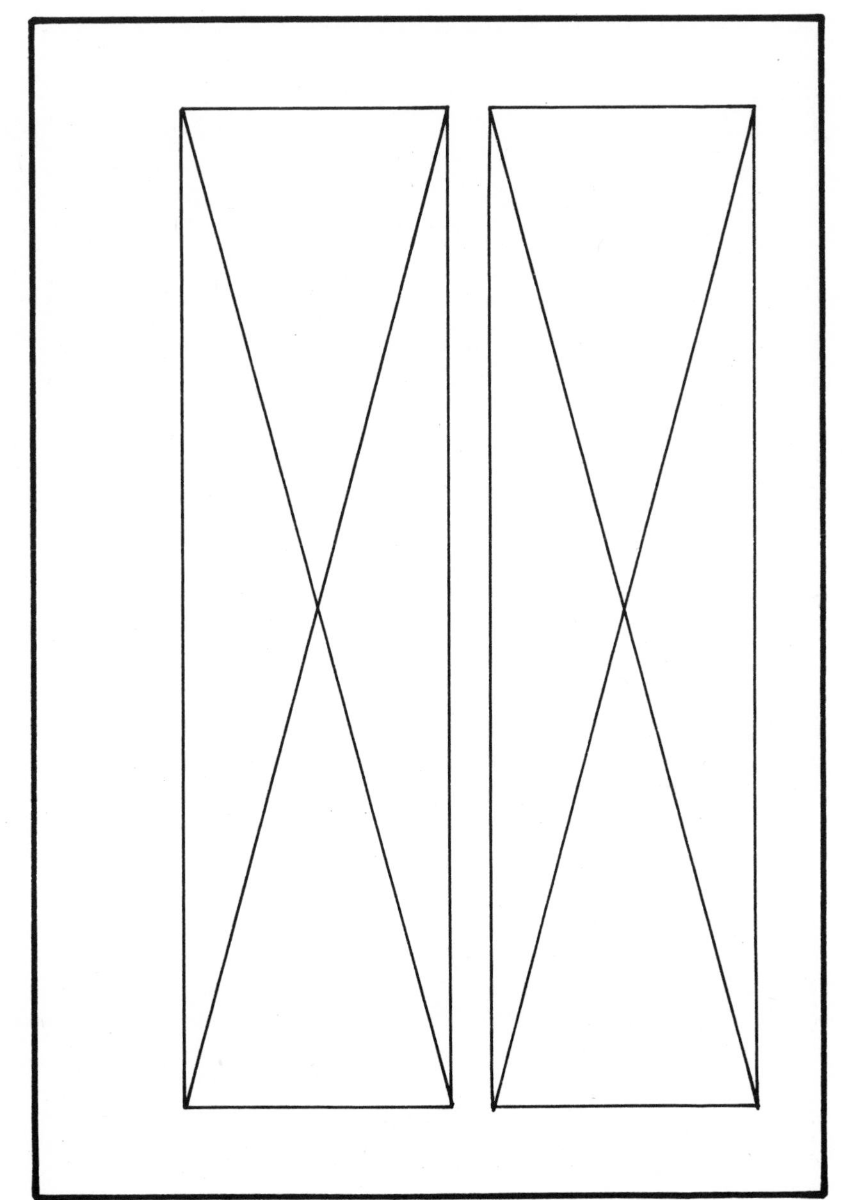

Figure 2-49.

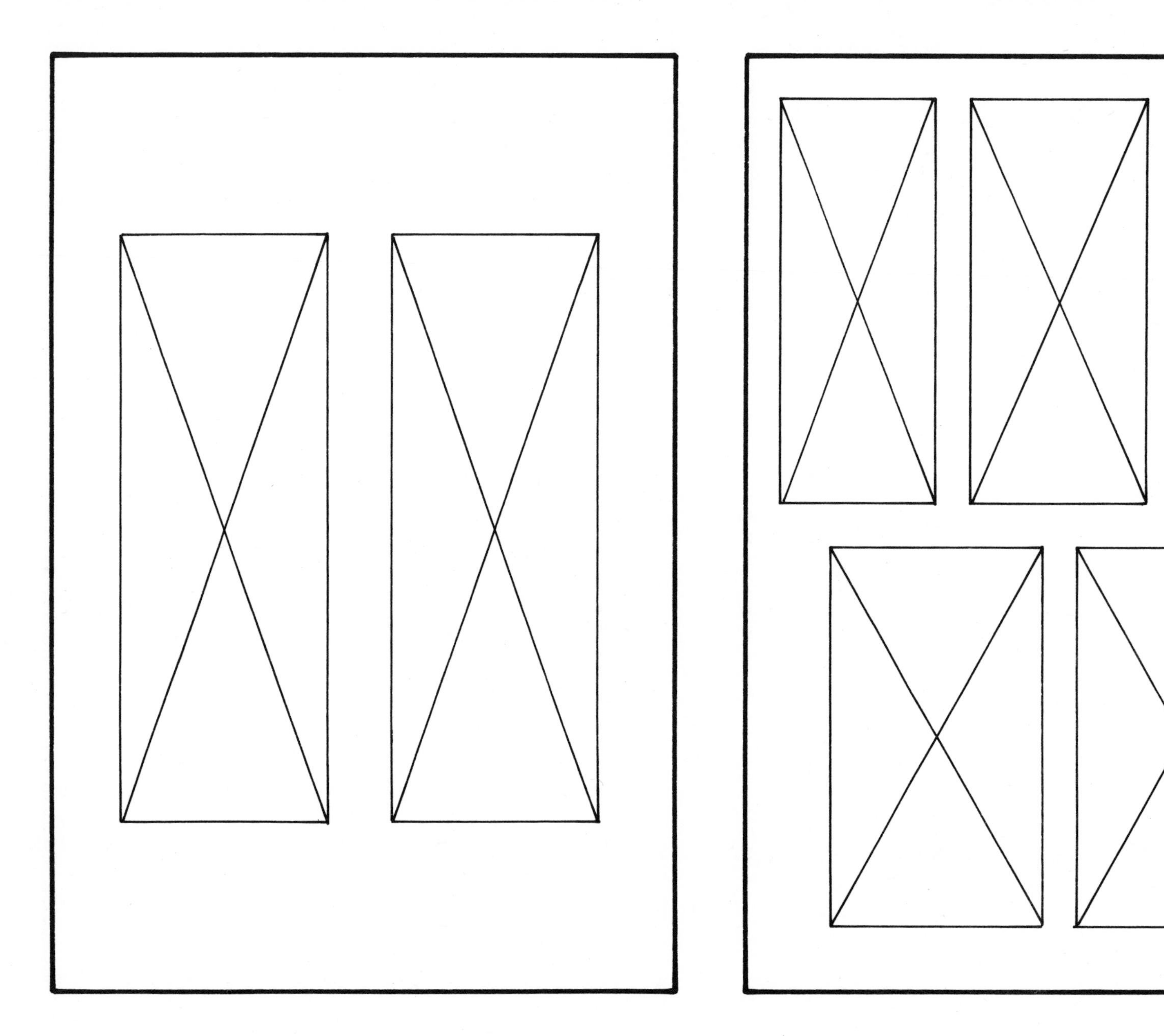

Figure 2-50.

Figure 2-51.

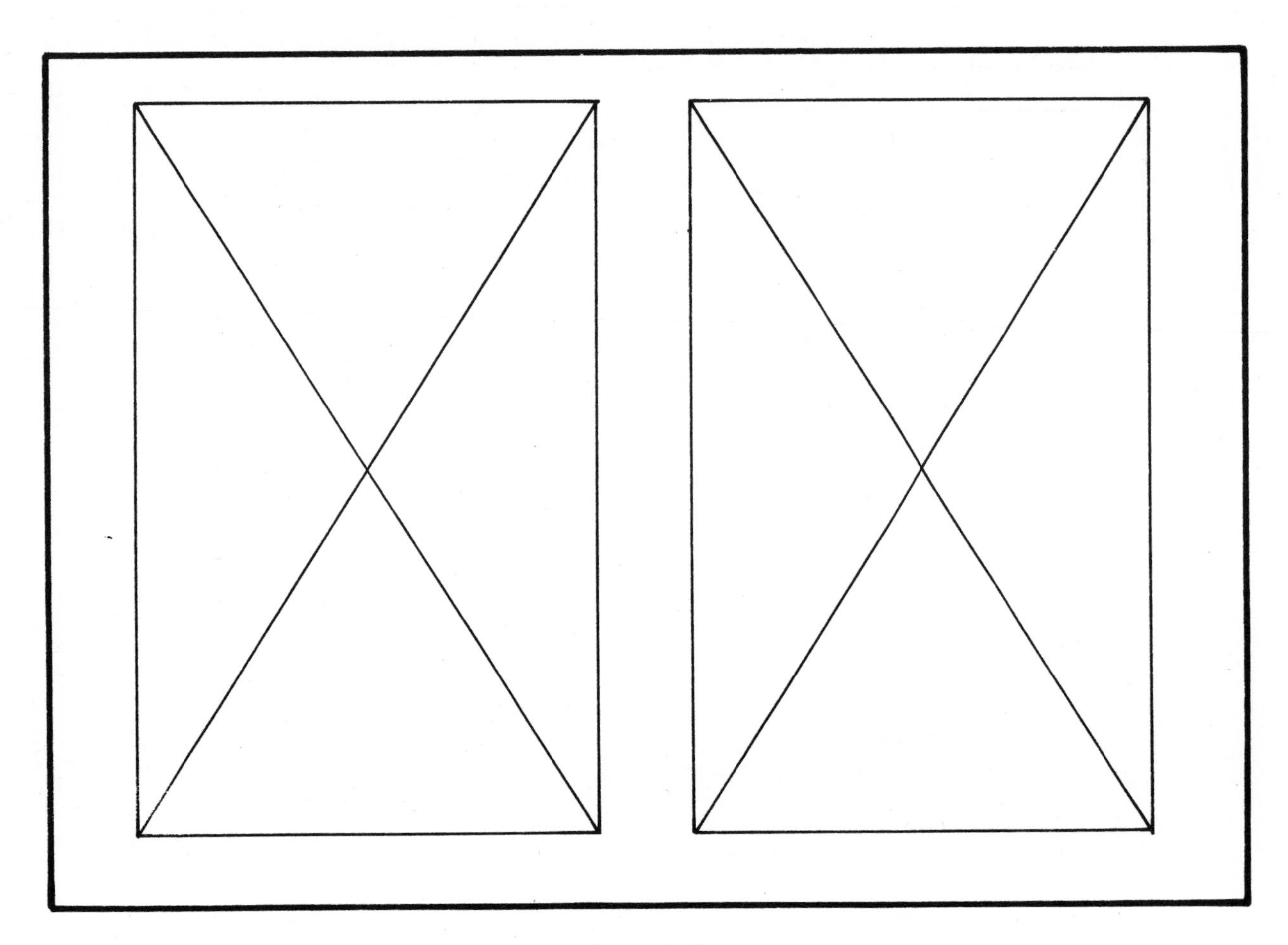

Figure 2-52.

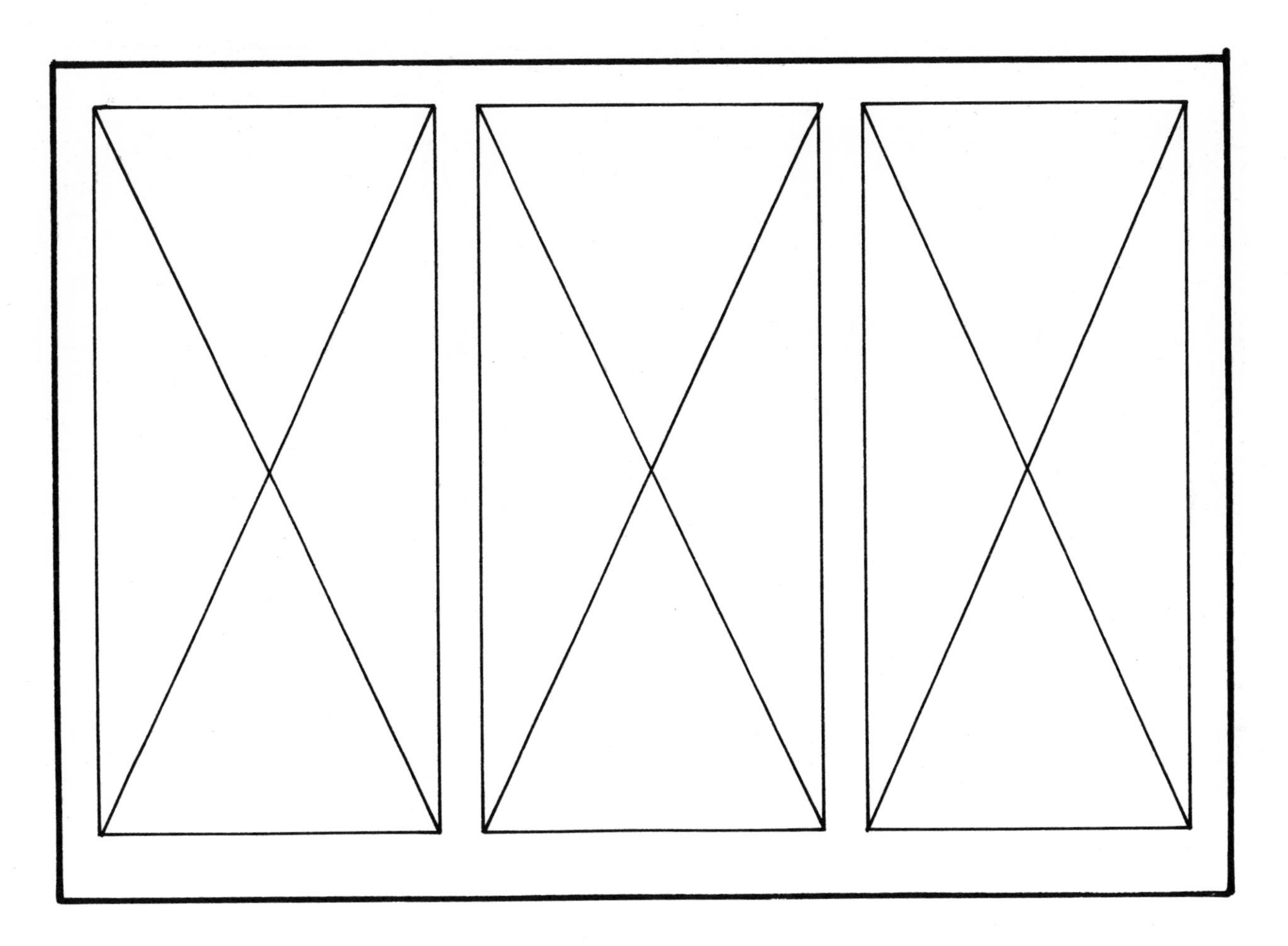

Figure 2-53.

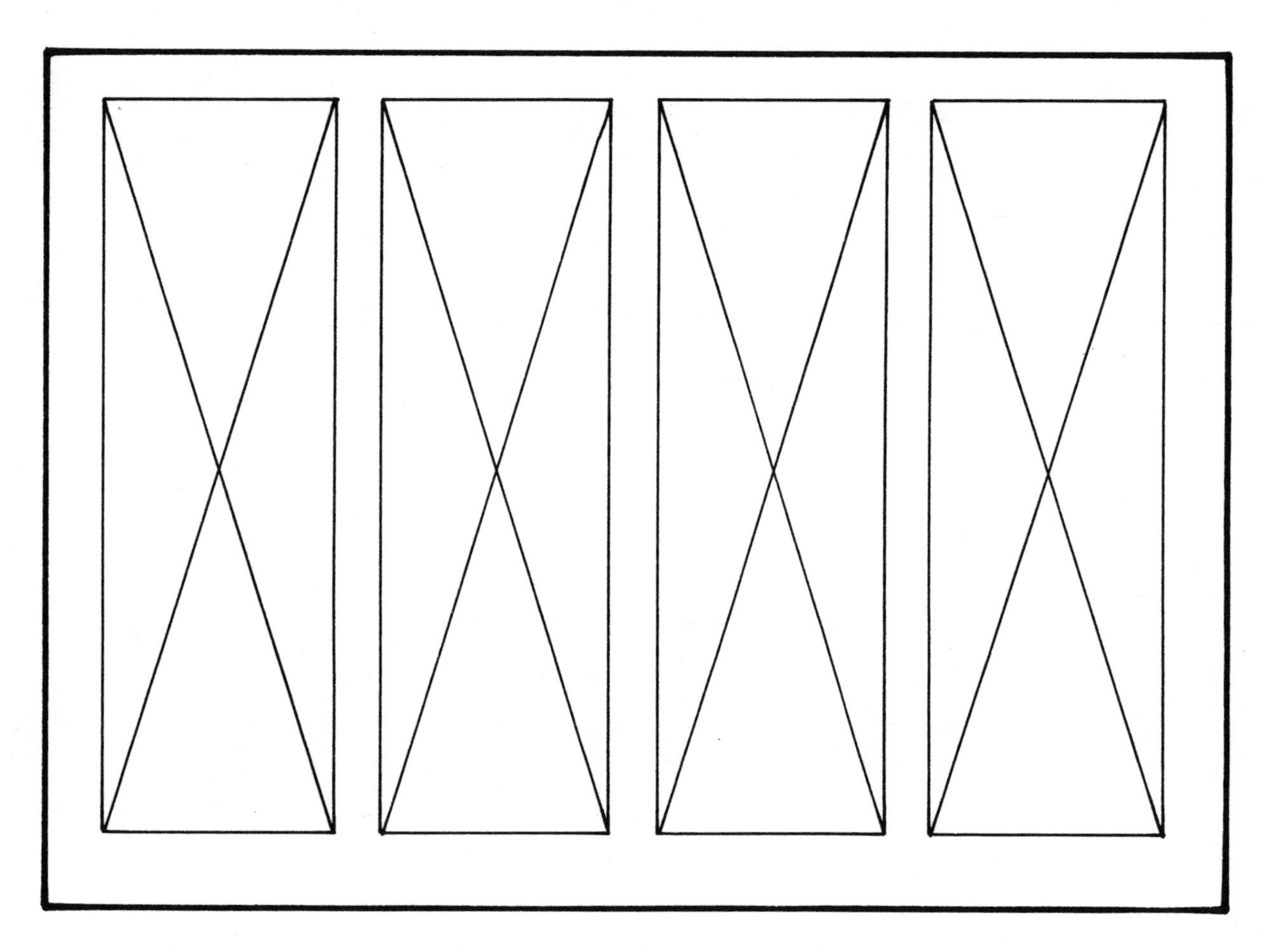

Figure 2-54.

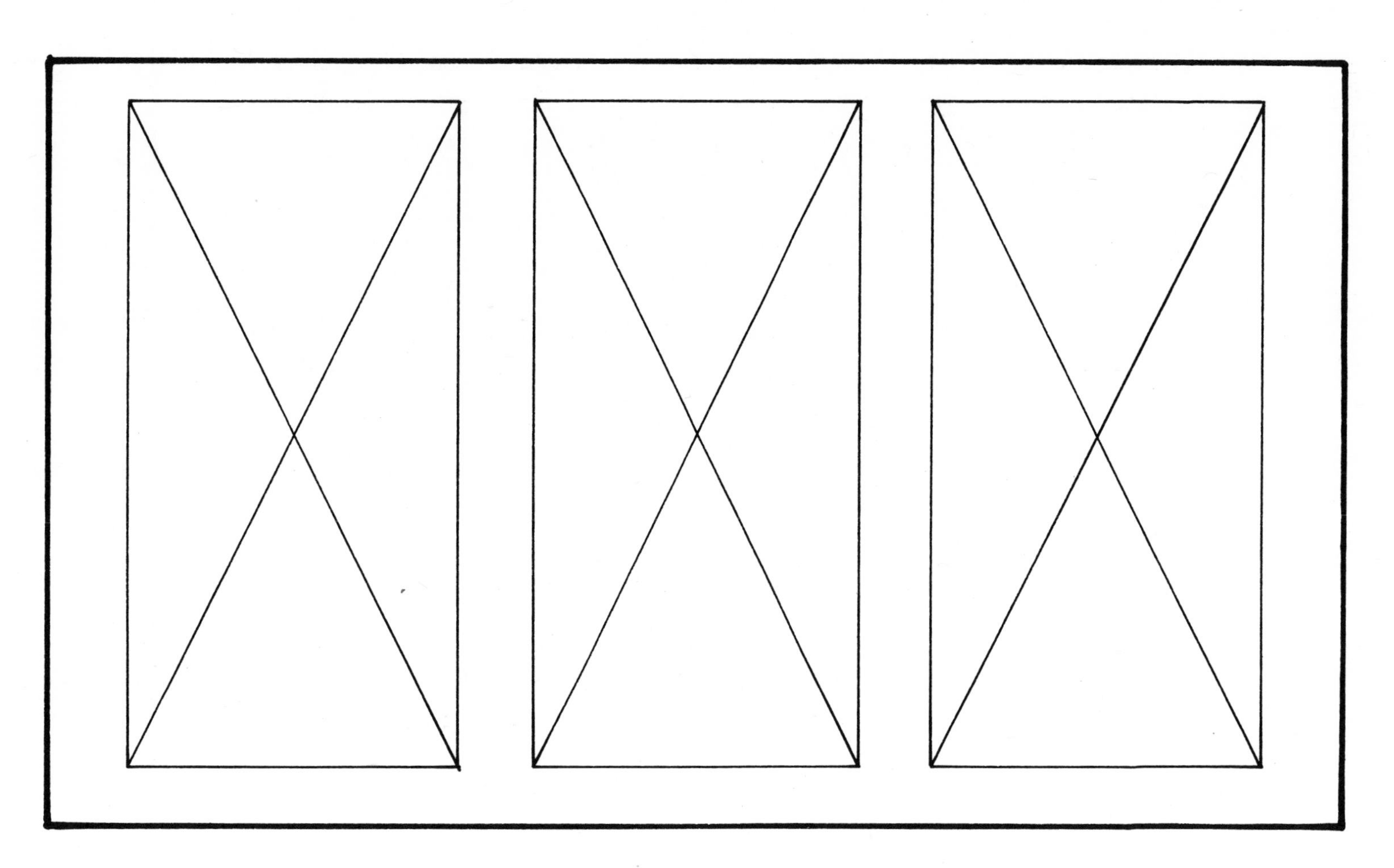

Figure 2-55.

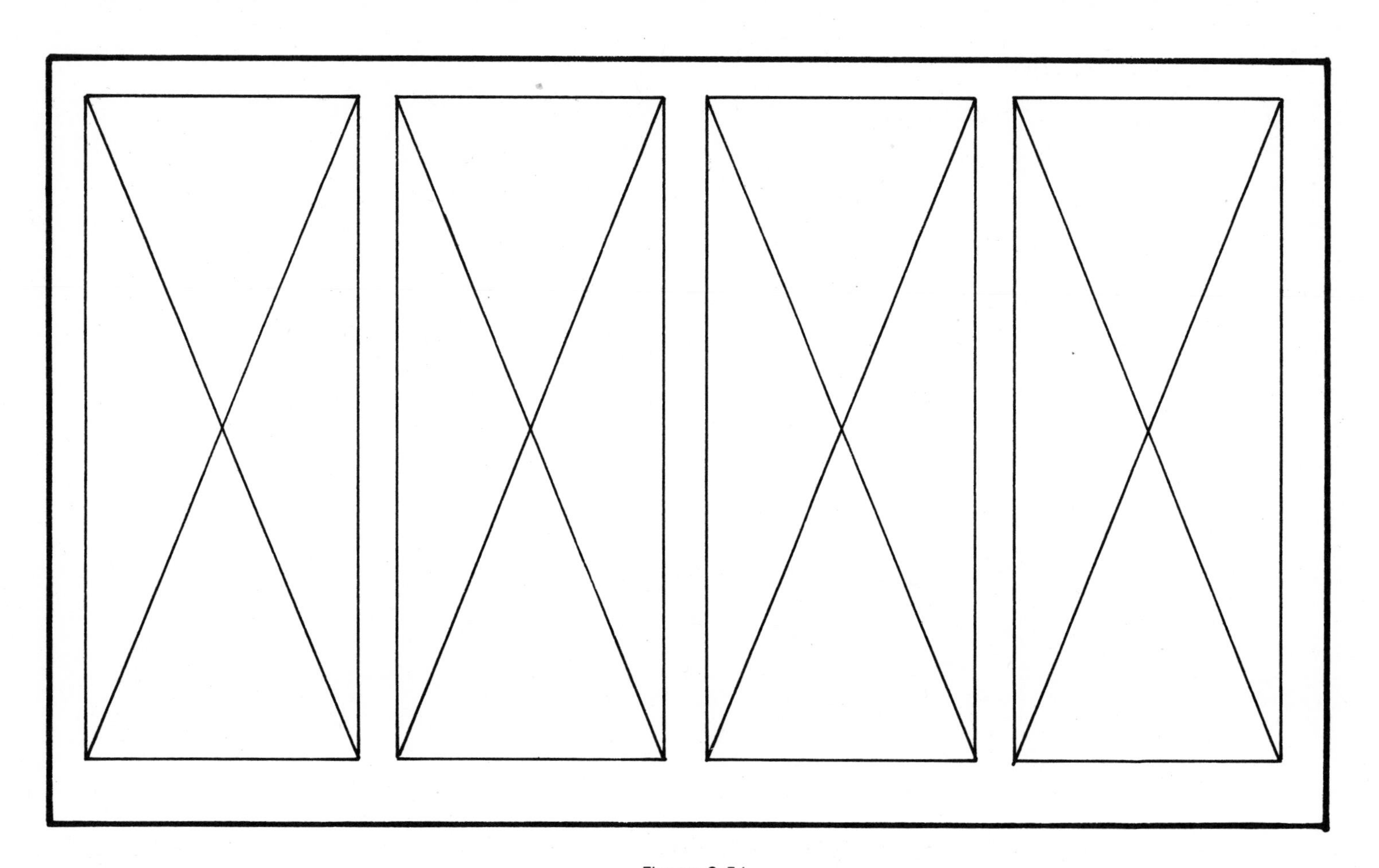

Figure 2-56.

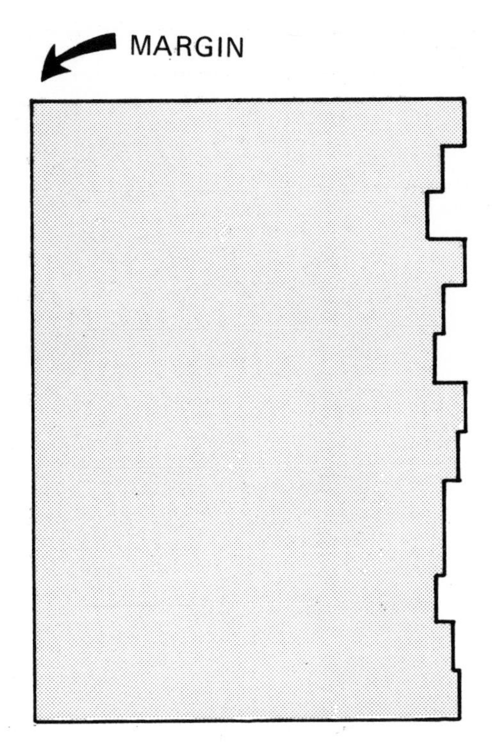

Figure 2-57.

Figure 2-58.

4. *Centered.* Both the right and left side of the text are ragged. It is very formal and is used more for short copy, invitations, or announcements. (Figure 2-60)
5. *Contoured.* This usually "fits" an accompanying line art or photograph. It can be interesting, but usually needs considerable text to be effective. (Figure 2-61)
6. *Runaround.* This margin wraps around graphics or photographs and is used in many books and magazines. (Figure 2-62)
7. *Shaped.* This technique can best be used to exaggerate eye movement across specific graphics or the overall page. Use small amounts of text because too much will create confusion and may be difficult to read. (Figure 2-63)

Designing with Type

Typography involves the setting and arranging of types and printing from them. Like handwriting, the style of the type chosen for a report will determine the character and quality of the finished product. No other component plays such an important role in the readability of the composition. (Figures 2-64 and 2-65)

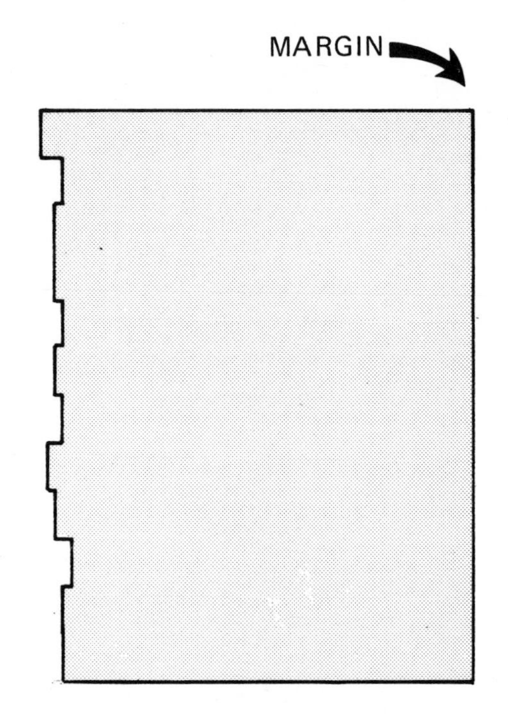

Figure 2-59.

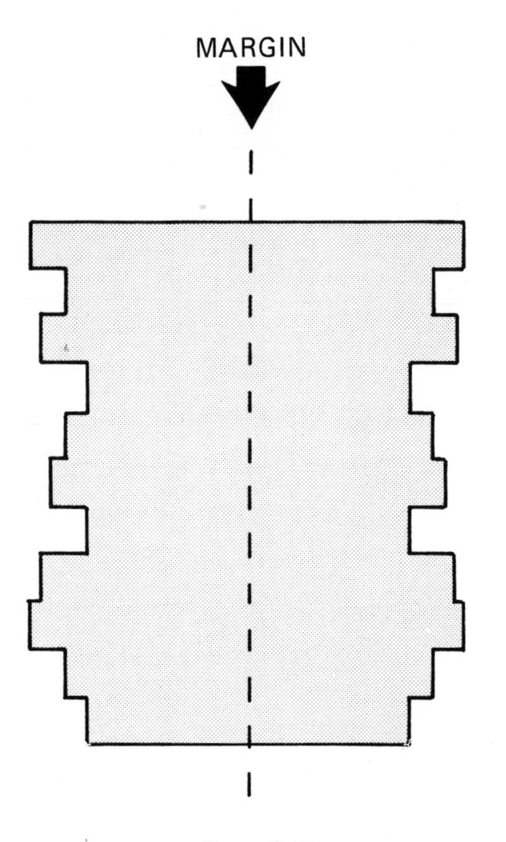

Figure 2-60.

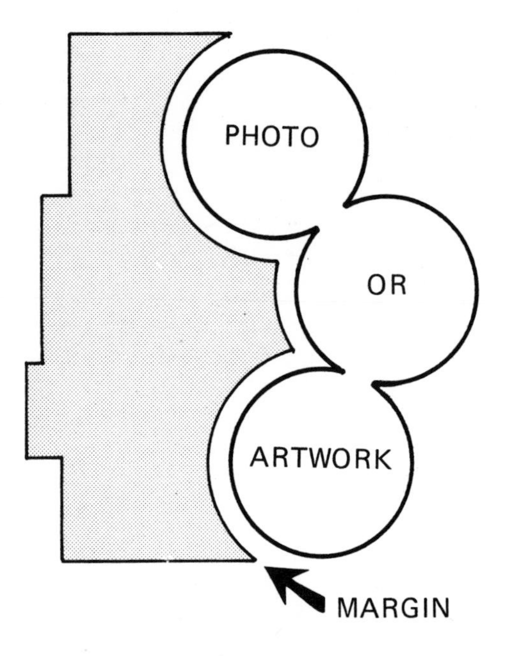

Figure 2-61.

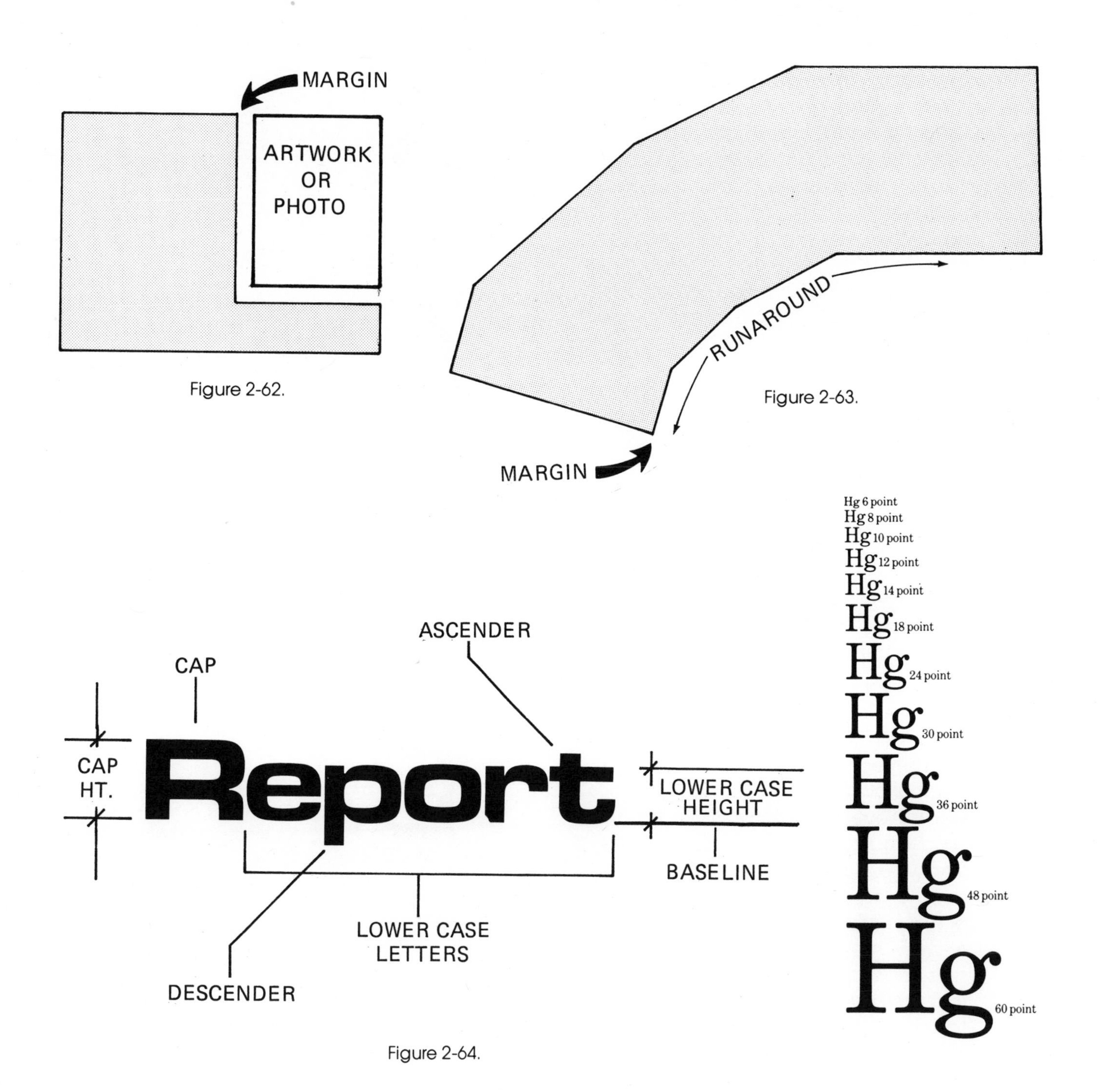

Figure 2-62.

Figure 2-63.

Figure 2-64.

There are three main categories of type forms, with unlimited variations of each. First, there is the *old style,* which is characterized by thick slanting serifs and the general appearance of being handwritten. Second, there is the *modern* style, which has a strong contrast between the thick and thin strokes. Its serifs are always crisp and clean. Third, there is the *contemporary* style, which includes all types without serifs, or sans serif letters. There is little difference between the strokes, but when they do change, they are referred to as "light," "medium," or "bold." (Figure 2-66)

Contrast is important in developing readability and legibility for type styles. Readability is the arrangement of the type, while legibility is the speed with which it can be recognized. Visual conflicts may arise if styles are mixed at random throughout the text, especially in titles or headings. It is important to avoid the following uses of type (Figure 2-67):

1. *Small and large.* Capitals and lower case letters of the same point size (type size) are acceptable, but different capitals or lower case letters of different points are not recommended.
2. *Thick and thin.* One is strong, the other is weak. Never mix the two styles together.
3. *Solid and outlined.* This is a matter of opposite contrast. They conflict with one another and, possibly, with the remainder of the page.
4. *Narrow and wide.* The horizontal emphasis is important to each style. It is difficult for narrow and wide type to work together.
5. *Hard and soft.* One style has defined edges, the other does not.

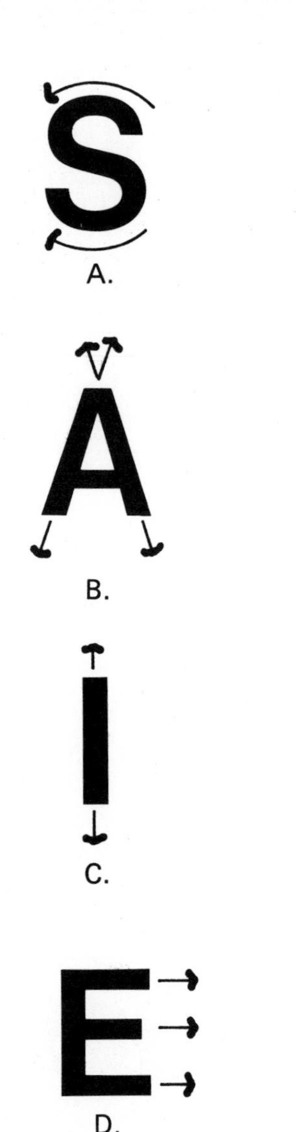

Figure 2-65. Letterforms are made up of four kinds of strokes: A = Cursive, B = Inclined, C = Vertical, D = Horizontal.

Figure 2-66.

SMALL/LARGE

THICK/THIN

HARD/SOFT

SOLID/OUTLINE

NARROW/WIDE

Figure 2-67.

Type is measured vertically by a point system. The larger the point number, the taller the type. Text sizes are usually no larger than 12 points, while headings can be as large as 42 points. Main titles are not restricted to a specific size, but page composition should guide its selection. (Figure 2-68 through 2-72)

The spacing of the type will be guided by the factor of optical judgment. Try to create a consistent volume of space between each letter. Vertical letter strokes are usually easier to space because there is less volume between them. Inclined strokes have the most volume and may be placed closer together. Cursive is midway between the other two strokes and can be more easily adjusted. (Figures 2-73 and 2-74)

THE SUPPORTIVE MATERIALS

There are basically four channels of expression in supportive report graphics: (1) photographs, (2) line art illustrations, (3) symbolic art, and (4) typography. Each has its place in the final product, but photographs and line art are used extensively in most design reports. Symbolic art includes the various logos that are developed by firms for clients, while typography is the technique for printing the written narrative.

Photographs

Practically any subject can be represented by a photograph. The only limitations are those that are self-imposed. Even line drawings can be photographed in black and white or color and placed on a page with special type or line art. Several photographs can be superimposed to create a multi-image pattern for special effect.

When using photographs in a report, the angle of the camera is critical. The high-angle

Instant Lettering® Standard Range

These pages show all the typefaces available on Instant Lettering sheets. They are grouped into Sans Serif, Serif and Decorative categories to help you select the right style. The full reference is shown on the page indicated by each typeface.

Sans Serif

Pg.	Typeface
18	Alternate Gothic No. 2
22	Antique Olive MEDIUM
22	Antique Olive SEMI BOLD
22	Antique Olive BOLD
24	Antique Olive BOLD CONDENSED
24	Antique OLIVE NORD
24	Antique OLIVE COMPACT
24	ITC Avant Garde GOTHIC EXTRA LIGHT
26	ITC Avant Garde GOTHIC MEDIUM
26	ITC Avant Garde GOTHIC MEDIUM CONDENSED
26	ITC Avant Garde GOTHIC BOLD
26	ITC Avant Garde GOTHIC BOLD CONDENSED
38	Cable LIGHT
38	Cable HEAVY
40	Cable ULTRA HEAVY
52	Compacta LIGHT
52	Compacta
52	Compacta ITALIC
54	Compacta BOLD
54	COMPACTA OUTLINE
54	COMPACTA BOLD OUTLINE
56	Compacta BLACK
58	Corinthian LIGHT
58	Corinthian MEDIUM
58	Corinthian BOLD
58	Corinthian EXTRA BOLD
60	Din 16 m
60	Din 17 m
64	ENGINEERING STANDARD
64	eurostile MEDIUM EXTENDED
64	Eurostile BOLD
64	eurostile BOLD EXTENDED
66	Folio LIGHT
66	Folio MEDIUM
66	Folio MEDIUM EXTENDED
66	Folio BOLD
66	Folio BOLD CONDENSED
68	Folio EXTRA BOLD
68	Franklin Gothic
68	Franklin Gothic CONDENSED
70	Franklin Gothic EXTRA CONDENSED
70	Franklin Gothic BOLD
72	Futura LIGHT
72	Futura MEDIUM
72	Futura MEDIUM ITALIC
74	Futura DEMI BOLD
72	Futura BOLD CONDENSED
74	Futura BOLD
74	Futura BOLD ITALIC
74	Futura EXTRA BOLD
76	Futura EXTRA BOLD CONDENSED
76	Futura BLACK
76	Futura DISPLAY
76	Gill EXTRA BOLD
76	Gill EXTRA BOLD CONDENSED
78	Gill EXTRA BOLD OUTLINE
78	Gill Kayo
78	Gill Kayo CONDENSED
78	Gill Sans LIGHT
78	Gill Sans
78	Gill Sans BOLD
80	Gill Sans BOLD CONDENSED
80	Grotesque 9
80	Grotesque 9 ITALIC
82	Grotesque 215
82	Grotesque 216
82	Helvetica EXTRA LIGHT
86	Helvetica LIGHT
84	Helvetica LIGHT CONDENSED
84	Helvetica LIGHT ITALIC
87	Helvetica MEDIUM
84	Helvetica MEDIUM CONDENSED
88	Helvetica MEDIUM EXTENDED
84	Helvetica MEDIUM ITALIC
90	Helvetica MEDIUM OUTLINE
90	Helvetica MEDIUM REVERSED
88	Helvetica BOLD
88	Helvetica BOLD CONDENSED
88	Helvetica BOLD ITALIC
90	Helvetica EXTRA BOLD
90	Helvetica COMPACT
92	Horatio LIGHT
92	Horatio MEDIUM
92	Horatio BOLD
96	MICROGRAMMA MEDIUM EXTENDED
98	MICROGRAMMA BOLD EXTENDED
100	News Gothic
100	News Gothic CONDENSED
100	News Gothic BOLD
108	Pump LIGHT
110	Pump MEDIUM
110	Pump DEMI BOLD
110	Pump
110	Pump TRILINE
124	Univers 45
124	Univers 53
124	Univers 55
124	Univers 57
124	Univers 59
124	Univers 65
126	Univers 67
126	Univers 75
126	Venus BOLD EXTENDED

Figure 2-68. Courtesy, Letraset U.S.A.

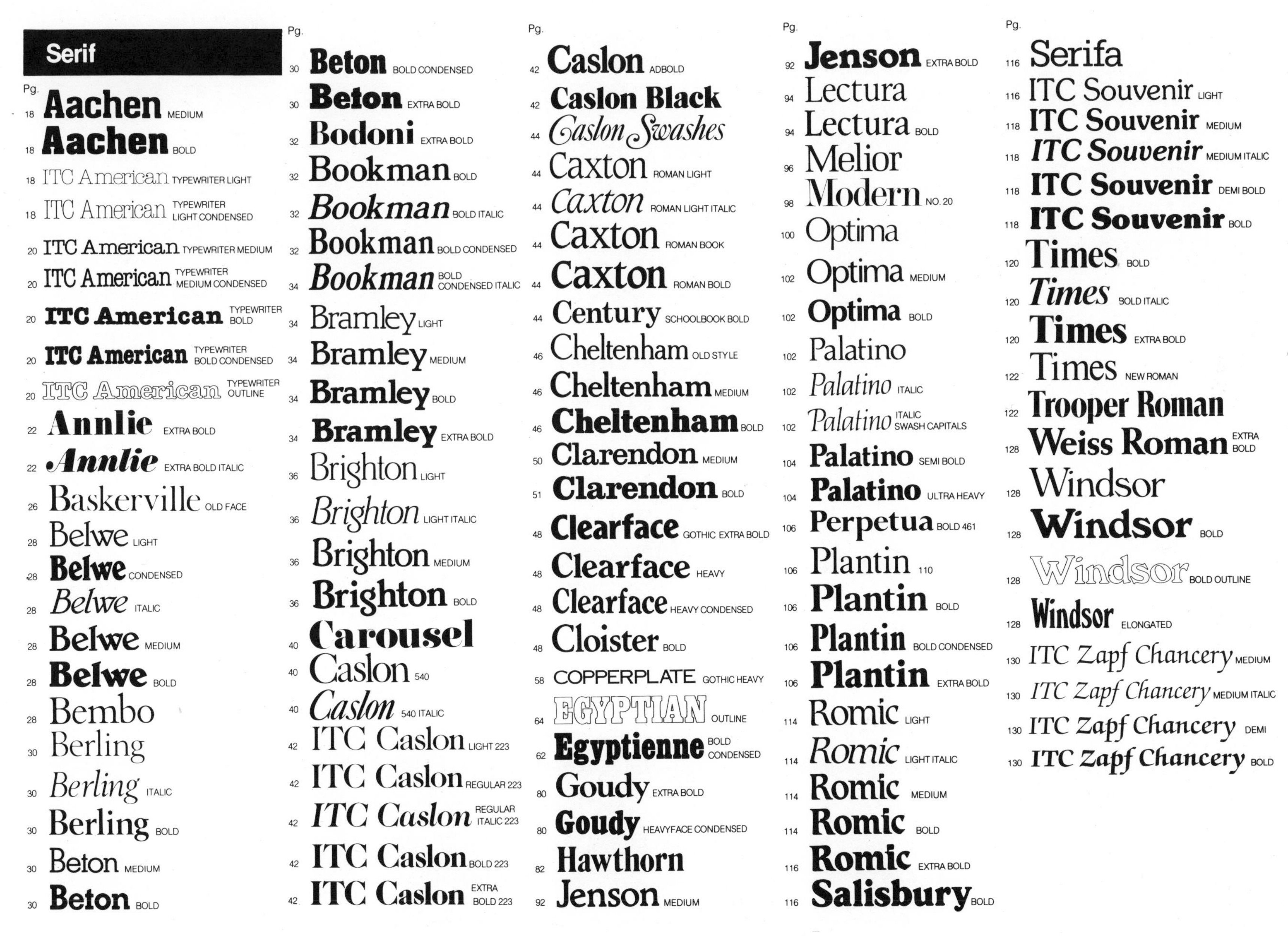

Figure 2-69. Courtesy, Letraset U.S.A.

Decorative

Pg.	Typeface
22	american uncial
24	Arnold Bocklin
32	Blanchard Solid
34	Bottleneck
36	Broadway
38	BROADWAY ENGRAVED
38	Brody
38	Brush Script
38	Bulletin TYPEWRITER
40	Candice
40	Candice Inline
46	CHARRETTE
46	City LIGHT
48	City MEDIUM
48	City BOLD
56	Cooper Black
56	Cooper Black ITALIC
56	Cooper Black OUTLINE
58	Countdown
60	Data 70
60	DAVIDA
60	DeVinne ORNAMENTED
60	Dom Casual
62	Dynamo MEDIUM
62	Dynamo
62	Dynamo CONDENSED
62	Dynamo SHADOW
62	Eckmann Schrift
64	Flash LIGHT
66	Flash
68	Fraktur BOLD
70	Freestyle Script
76	GALLIA
80	Goudy Fancy
82	Highlight
90	Hobo
92	Juliet
94	Kalligraphia
94	Lazybones
94	L.C.D.
94	Le Griffe
96	LETTRES ORNEES
96	Loose New Roman
96	MANUSCRIPT CAPITALS
98	Murray Hill Bold
100	Old English
102	Palace Script
104	Park Avenue
104	PEIGNOT LIGHT
104	PEIGNOT MEDIUM
104	PEIGNOT BOLD
106	Pendry Script
108	Playbill
108	Pretorian
108	PRISMA
108	PROFIL
110	QUENTIN
112	Revue
112	Ringlet
112	Rockwell LIGHT 390
112	Rockwell 371
112	Rockwell BOLD 391
114	Rockwell EXTRA BOLD
114	ROMANTIQUES NO. 5
116	SANS SERIF SHADED
116	SAPPHIRE
118	Stencil Bold
118	Tabasco MEDIUM
120	Tabasco BOLD
122	Tintopetto
122	Tip Top
126	University Roman
126	University Roman BOLD
128	Vivaldi
130	Zipper

Figure 2-70. Courtesy, Letraset U.S.A.

White Instant Lettering®

Sans Serif

Pg	Typeface
22	Antique Olive MEDIUM
22	Antique Olive SEMI BOLD
22	Antique Olive BOLD
24	Antique Olive BOLD CONDENSED
24	Antique OLIVE NORD
24	Antique OLIVE COMPACT
26	ITC Avant Garde GOTHIC MEDIUM
26	ITC Avant Garde GOTHIC MEDIUM CONDENSED
26	ITC Avant Garde GOTHIC BOLD
26	ITC Avant Garde GOTHIC BOLD CONDENSED
38	Cable HEAVY
40	Cable ULTRA HEAVY
56	Compacta BLACK
58	Corinthian MEDIUM
58	Corinthian BOLD
58	Corinthian EXTRA BOLD
64	eurostile BOLD EXTENDED
66	Folio MEDIUM
68	Franklin Gothic
68	Franklin Gothic CONDENSED
70	Franklin Gothic EXTRA CONDENSED
72	Futura MEDIUM
74	Futura DEMI BOLD
72	Futura BOLD CONDENSED
74	Futura BOLD
76	Futura EXTRA BOLD CONDENSED
78	Gill Kayo
78	Gill Kayo CONDENSED
86	Helvetica LIGHT
84	Helvetica LIGHT CONDENSED
87	Helvetica MEDIUM
84	Helvetica MEDIUM CONDENSED
88	Helvetica MEDIUM EXTENDED
84	Helvetica MEDIUM ITALIC
90	Helvetica MEDIUM OUTLINE
90	Helvetica MEDIUM REVERSED
88	Helvetica BOLD
88	Helvetica BOLD CONDENSED
88	Helvetica BOLD ITALIC
90	Helvetica COMPACT
90	Helvetica EXTRA BOLD
98	MICROGRAMMA BOLD EXTENDED
108	Pump LIGHT
110	Pump MEDIUM
110	Pump DEMI BOLD
110	Pump
110	Pump TRILINE
124	Univers 65
124	Univers 67
124	Univers 75

Serif

Pg	Typeface
18	Aachen MEDIUM
18	Aachen BOLD
20	ITC American TYPEWRITER MEDIUM
20	ITC American TYPEWRITER MEDIUM CONDENSED
20	ITC American TYPEWRITER BOLD
20	ITC American TYPEWRITER BOLD CONDENSED
28	Belwe LIGHT
28	Belwe CONDENSED
28	Belwe MEDIUM
28	Belwe BOLD
28	Bembo
30	Beton BOLD CONDENSED
30	Beton EXTRA BOLD
32	Bodoni EXTRA BOLD
32	Bookman BOLD
32	Bookman BOLD ITALIC
32	Bookman BOLD CONDENSE
34	Bookman BOLD CONDENSED ITAL
34	Bramley LIGHT
34	Bramley MEDIUM
34	Bramley BOLD
34	Bramley EXTRA BOLD
36	Brighton LIGHT
36	Brighton LIGHT ITALIC
36	Brighton MEDIUM
36	Brighton BOLD
42	ITC Caslon LIGHT 22
42	ITC Caslon REGULAR

Figure 2-71. Courtesy, Letraset U.S.A.

Figure 2-72. Courtesy, Letraset U.S.A.

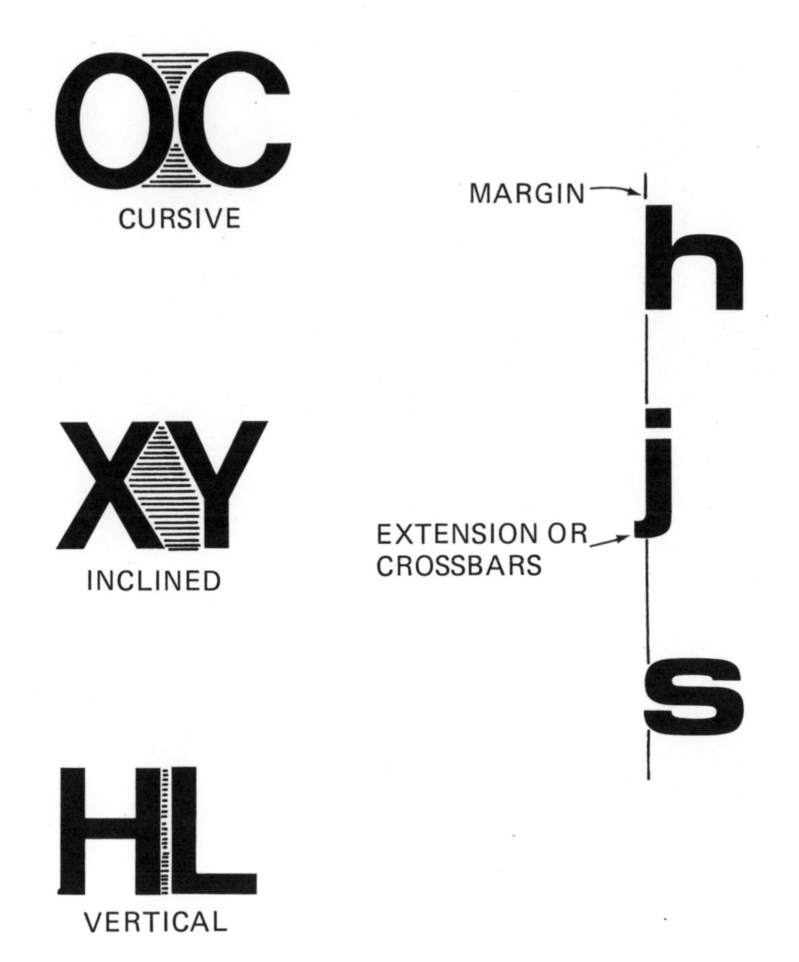

STRAIGHT, VERTICAL
OR INCLINED

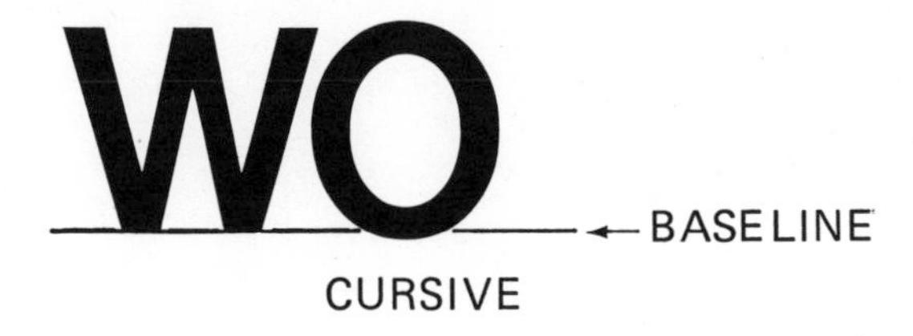

Figure 2-73. The spacing and placement of the letters in the title or heading must be optically correct to be understood by the reader.

A. The Title of the Report

BASELINE #1

BASELINE #2

#3

Figure 2-74. For a comfortable reading quality for report titles, be careful of the spacing at the baselines. Title *A* is measured from the base of the letters. Title *B* is measured to allow space for the ascenders and descenders.

exposure is best for representing important relationships between structure and site or the arrangement of space. A medium angle is better if a structure or site feature must be individually emphasized. The low-angle shot is better if an individual component within the element must be represented. (Figure 2-75)

Heat-sensitive screens can also be applied to some photographs for greater visual impact. This technique will allow a report designer to draw attention to or away from special areas of a page. Grid, dot, or line patterns are the most popular, but the patterns chosen are limited only by the individual imagination. (Figures 2-76 through 2-79)

Line Art

Of the supportive graphics, line art is the most important in a design report. It includes action sketches, perspective drawings, mapping components, construction details, graphs, charts, and flow diagrams. Their use is important to support the written composition and to increase readability.

Report designers, however, should avoid abstract illustrations that might confuse the reader. The written text is the most important element in the report, and line art should be used only for its support. Line illustrations that do not support the text are referred to as eye wash graphics. They are costly to produce and rarely benefit the final document.

Symbolic Art

This form of graphics has recently become very popular because of its instant readability. Traffic symbols and corporate logos usually dominate

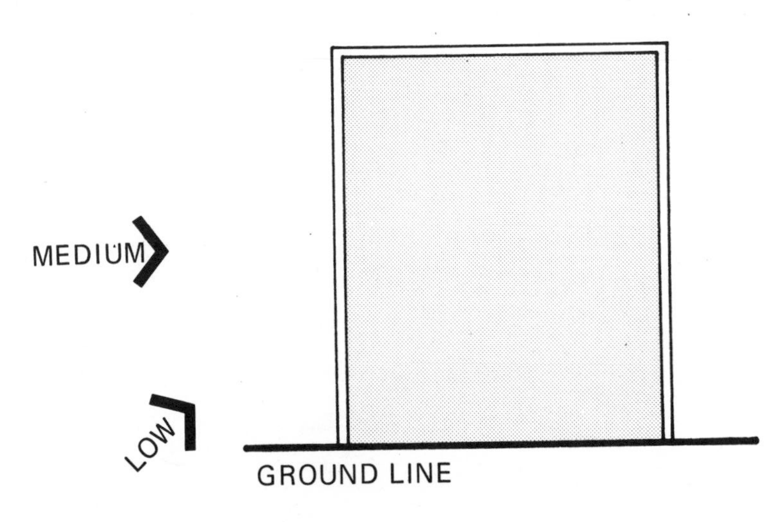

Figure 2-75.

this form. (Figure 2-80) A particular message or thought can be transferred to the mind of the reader more easily and rapidly without lengthy lines of type. Directions can be given, ideas communicated, and impressions made without extensive production costs.

Typography

The extensive use of artistic typography is not always a positive approach to report composition. Although printing technology has developed to the point that alphabet variations are unlimited, report type should remain simple and uncomplicated. This is not only because most

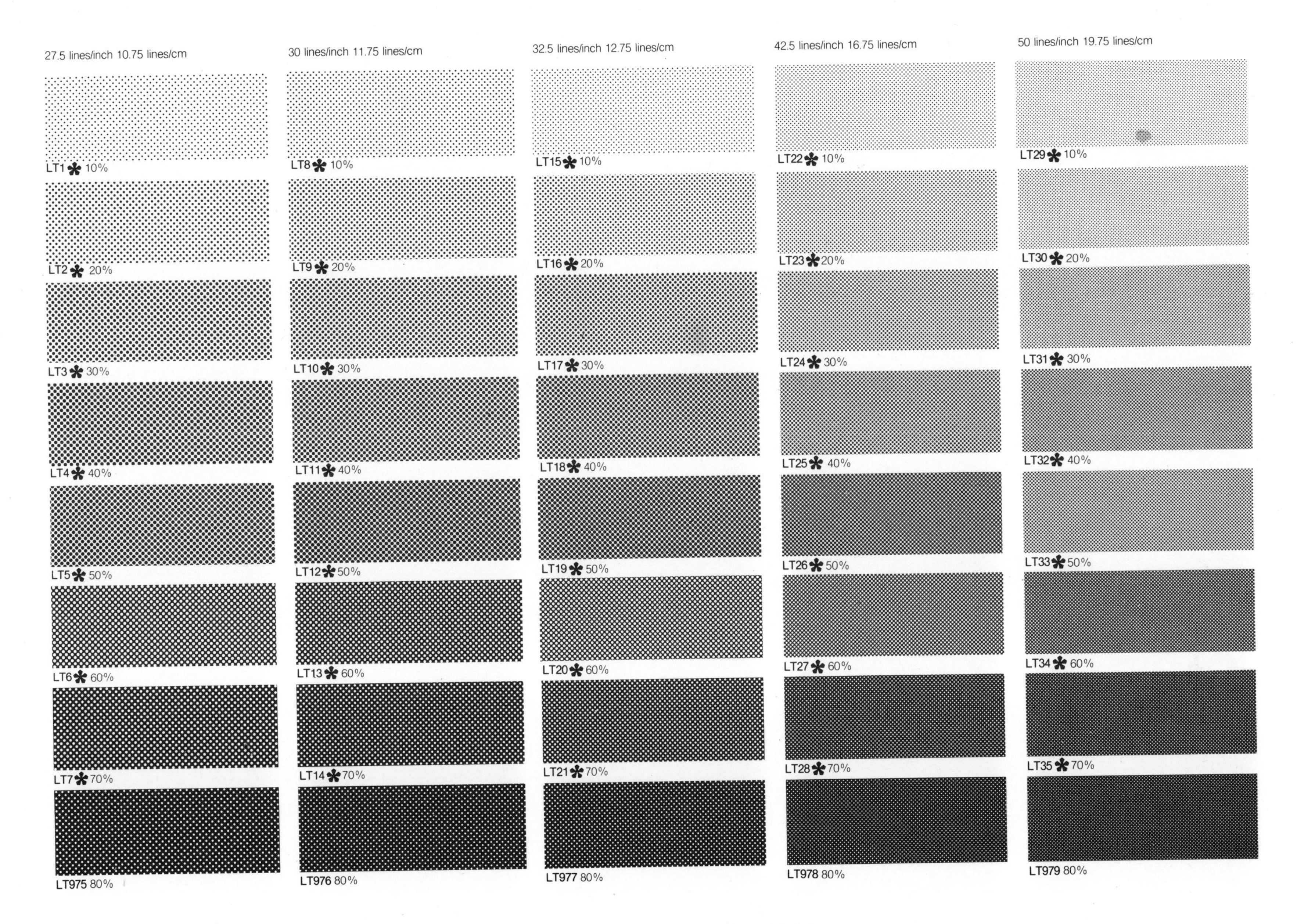

Figure 2-76. Courtesy, Letraset U.S.A.

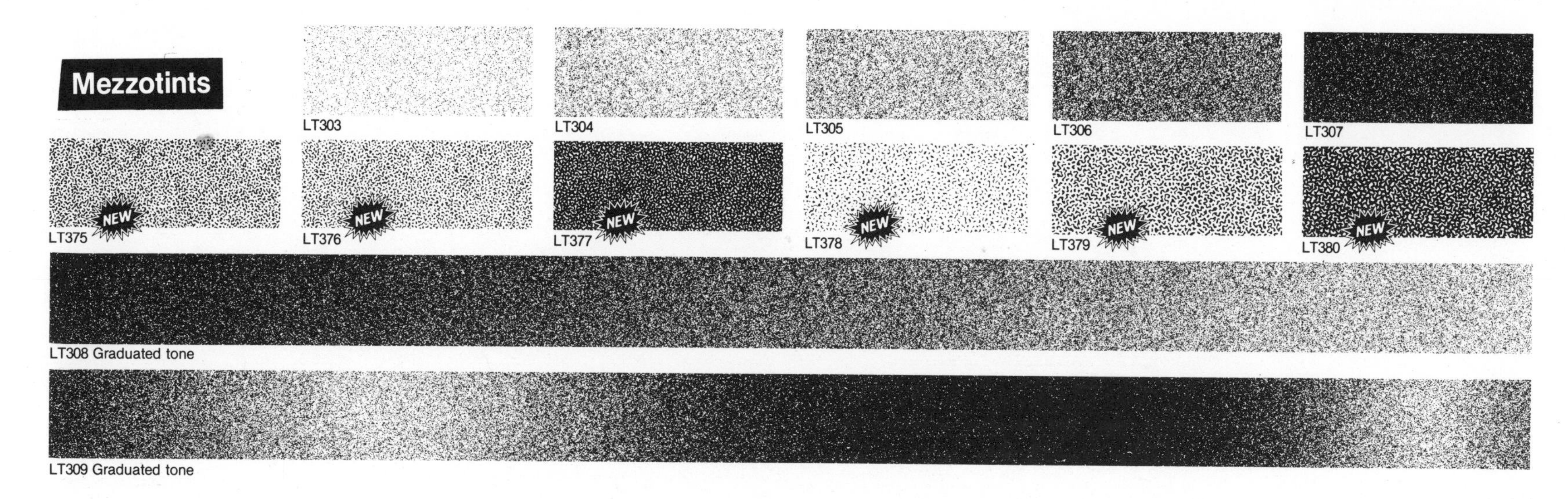

Figure 2-77. Courtesy, Letraset U.S.A.

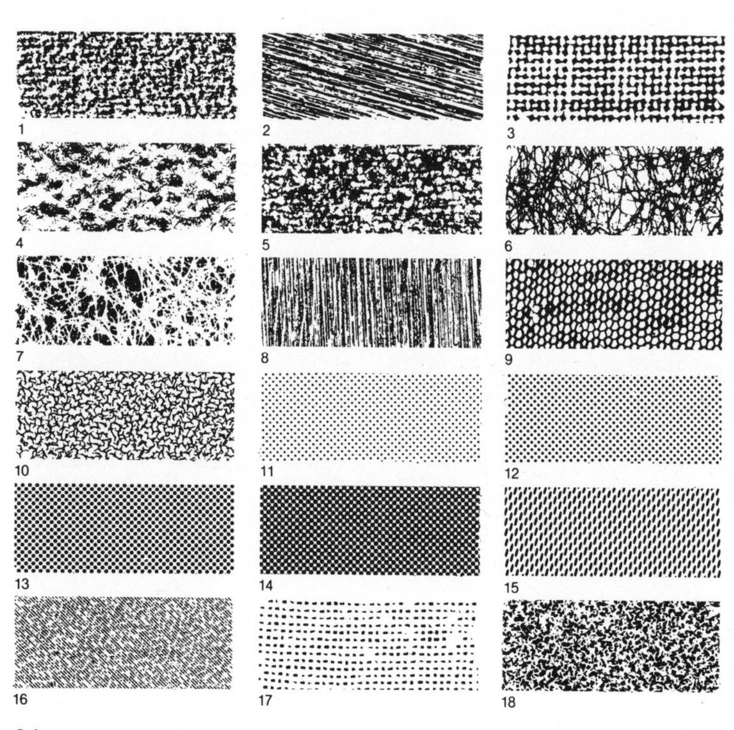

Figure 2-78. Courtesy, Letraset U.S.A.

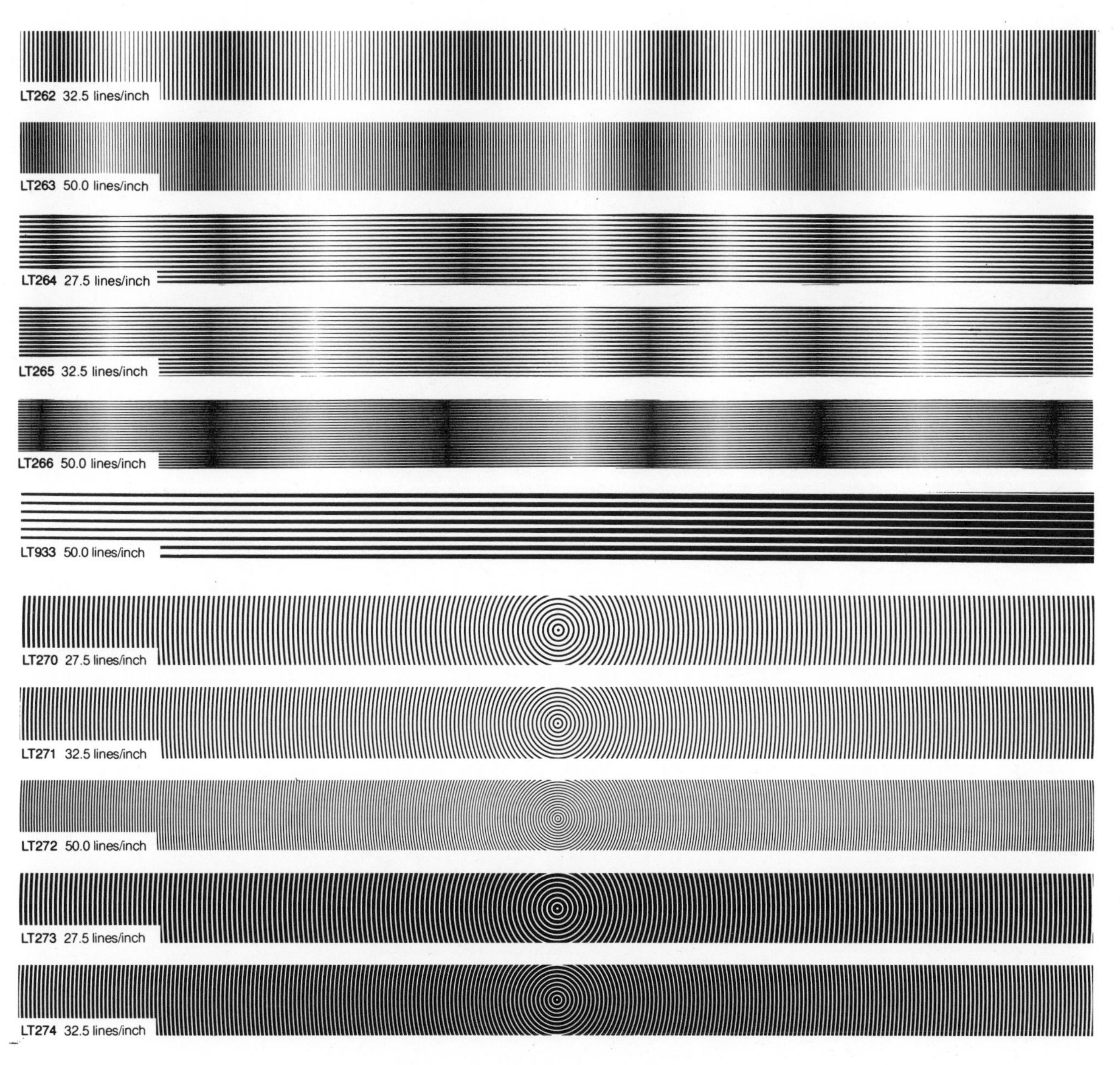

Figure 2-79. Courtesy, Letraset U.S.A.

Figure 2-80.

TONE
TONE
TONE
TONE
TONE

Figure 2-81.

readers prefer a less complicated type style when reading, but also because they are better able to visually decode the more simple lines of print.

Research by Huey (1968) and Tinker (1963, 1966) established, for example, that the more educated reader processes only part of what is seen. The more education a person has, the less he or she actually needs to see of each letter in order to identify it correctly. When reading left to right, the right-hand side of the letter is more informative. When reading top to bottom, the upper portion of the printed line is more informative. If any variations in type style are used, for example in titles or headings, these factors should be considered.

The Use of Color and Tones

It is important to remember that the costs of producing the report will rise if color is used. Several press runs will be required, and with each run there is an additional step in the production process. Color, therefore, should be used to emphasize an important aspect of the design and not primarily to decorate it. The following techniques, however, can be employed to enhance the effects of the final document:

1. Print the title or division headings in a second color.
2. Print the type in color, using a complimentary color for the paper stock. Then print the title and division headings in black.
3. Print the line art in a second or third color.
4. Print the various phases of a design project in different colors.
5. Use a three-color photograph for the cover, or use them throughout the text.

One-color reports can be toned or screened to create the illusion of color. This is not as expensive as multicolor processes but is often as effective. Figures 2-81 and 2-82 show examples of toning in design reports.

Making Reductions

Most designers are conditioned to using large sheets of paper to produce their details and line drawings, and they often forget that the average report page is only 8½ × 11 inches. Because of this, certain artwork is often reduced too much, making it difficult for the average reader to comprehend. The grid page is a convenient guide for determining the reduction or enlargement of supportive art. Just use the grid space you need (as developed in the preliminary layout, double the dimensions, complete the artwork, and reduce the finished art by 50 percent. This can be accomplished with most in-house copy machines, and expensive camera work can be avoided. (Figure 2-83)

Folding and Binding

There are numerous types of folding and binding techniques that can be used to organize the final report product. The basis for determining which to use should be the importance of the document, the audience, the convenience of use, and the report's life expectancy. See Appendix 1 for information on folding and Appendix 2 for binding.

The final appearance of the report page can be as simple or as complex as the report itself. Figures 2-84 through 2-111 show examples of numerous types of page layouts that can be produced by a report team.

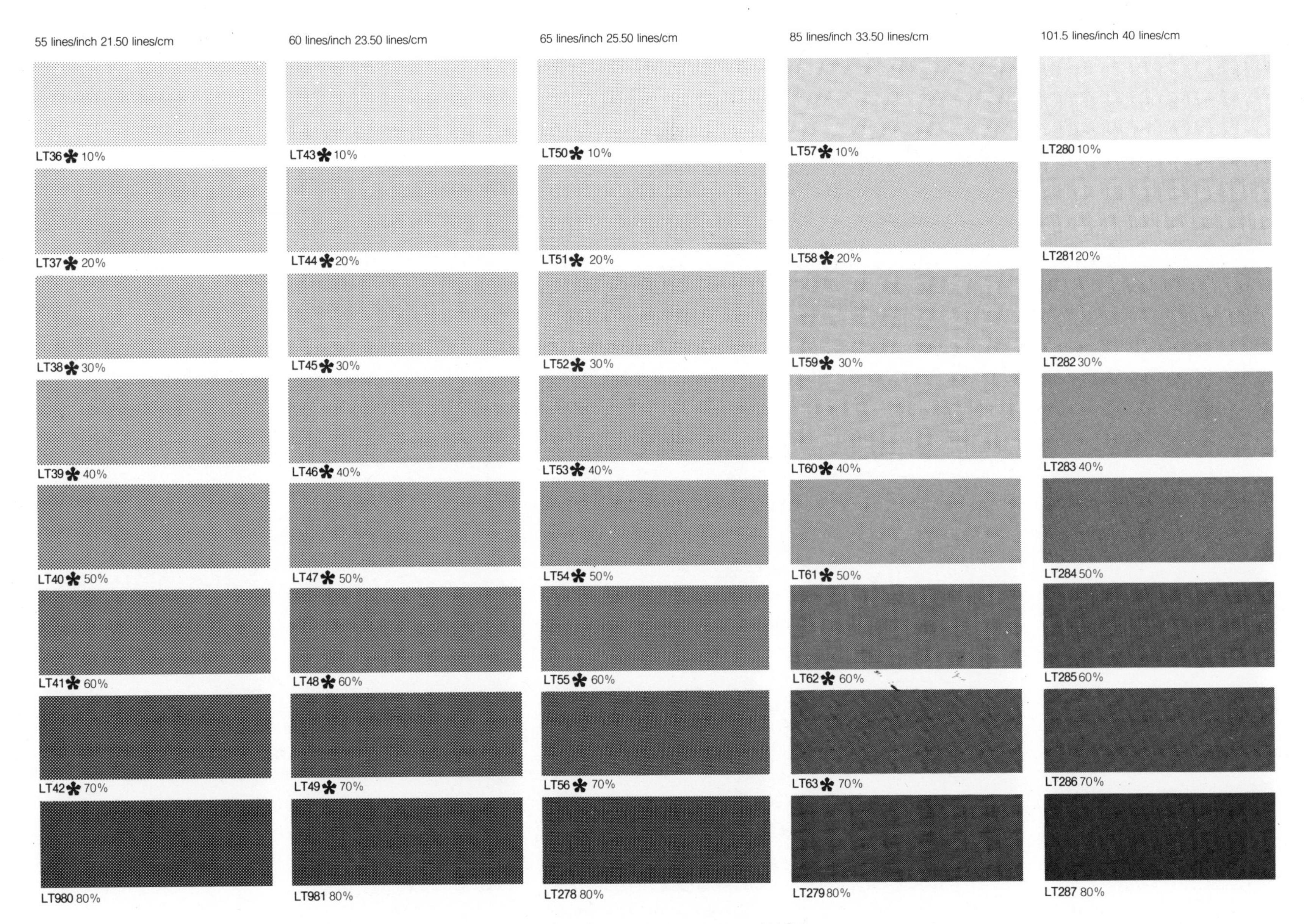

Figure 2-82. Courtesy, Letraset U.S.A.

Figure 2-83. Sketch *1* represents the typical size for a line drawing. Sketch *2* represents the same drawing reduced to "fit" within the printed document. Courtesy, Johnson, Johnson and Roy, Inc.

CLARITY FROM CONFUSION! CONSIDER ALL THE RESOURCES TO GET YOUR VOLUNTEER PROGRAM OFF THE GROUND.

Figure 2-84. Courtesy, U.S. Department of the Interior.

Landscape Character

Landscape development for the cemetery has carefully considered the retention of the natural setting, not the creation and maintenance of a manicured landscape.

Visitors will notice that care has been taken to conserve existing natural areas, woodlots, slopes and wetlands. Where construction of roads, buildings and burial sections has disturbed natural areas, trees, shrubs and ground cover native to this area have been used to revegetate disturbed areas. Native plants have been used to reforest some areas formerly cleared for farm fields and to ensure privacy within the cemetery by screening adjacent public roads.

Consistent with energy conservation needs and environmental responsibility is the necessity to understand techniques for *maintaining* the native landscape. The existing sandy loam soil in some areas of the cemetery is thin and impoverished. Any disturbance to the soil cover could result in permanent scars unless the soil were enriched and replanted. Plants and grass mixtures indigenous to this site are used in burial sections and in replanted areas since they will require minimal care.

A sensitive planting program responds to these qualities, seeking either to reinforce the plant community, modify it by introducing alternative species, or eradicate it and establish a new community. Which of these alternatives is pursued depends not only on the quality of the community but also on the treatment of the land unit in accordance with the Master Plan.

Design Criteria

Considerations of conservation and maintenance of the natural landscape led to the design criteria discussed in this section.

Preservation

The quality of existing plant communities is reinforced through the removal of invader species, scrub undergrowth, diseased and damaged specimens. Extending these efforts into projected phase development helps to keep future maintenance costs low by preventing growth of undesirable species.

Reforestation

Areas that are cleared during construction are replanted to encourage plant associations that develop naturally under those specific site conditions.

Transition

Plantings soften the edge between developed and natural sectors. Such plantings unify these areas as well as provide a protective buffer where the woodland edge has been cleared.

Definition of Spaces

Spaces for different functions are enclosed by plantings to define the space. Extensions of the existing wooded areas provide smaller spaces of a more intimate scale.

Views

Plantings direct views by framing interesting and attractive features such as ponds, kettleholes, or the flagpole display area. Visual screens of plant materials serve to close off undesired views to perimeter public roads.

Energy Conservation

Vegetation is implemented as an energy conservation measure, providing buildings with a protective wind buffer during the winter and shade in the summer. Snow accumulation is controlled through the use of wind channels formed by planting masses.

Accent

In areas of special interest and in pedestrian zones, plantings provide color, texture, form and scent to highlight and emphasize the special character of these places.

Visual Direction

Landmark plantings of notable height, mass and contrast assist visitors in movement on the site.

Existing Character of the Site

Open fields of former farms with a variety of fieldstone, and prominent woodland with gently sloping terrain suggest a palette of material that closely relates to these natural settings.

30

Figure 2-85. Courtesy, Johnson, Johnson and Roy, Inc.

The Northeast Sector has a majority of the "fair" and "poor" structures in the community. A visual survey would also classify this part of town as the oldest sector. Not all the houses in this sector are in need of improvement, several structures having been maintained to a "good" or "excellent" standard.

The South Sector has a few square blocks of "good" and "excellent" housing. This sector is the least developed area of town with many vacant lots. It should be noted that the federally financed low-rent housing in this area has not been evaluated but would be considered "good" or "excellent." There are many "fair" and "poor" structures in the northern half of this sector, just south of the railroad tracks.

In general, the housing conditions could be rated as fair with a large number of "excellent" as well as "poor" structures in existence. A conscientious program of housing rehabilitation should be undertaken by the City Council with the help and assistance of the Planning Commission and the Housing Authority. The responsibility of the Housing Authority to improve the living environment of all citizens of the community should not be minimized. The Planning Commission should assist them in their initiatives whenever possible.

Although the City of Deshler does not have a major clearance-type renewal problem, the facts indicate that in the near future an investigation into a rehabilitation-type Neighborhood Development Project would be desirable. Any area which can qualify for rehabilitation treatment is also eligible to receive the three-fourths Federal grant for spot-clearance of scattered substandard structures as well as for the construction and installation of all necessary public improvements and utilities, such as streets, curbs, gutters, sidewalks, water and sewer lines, etc.

9

Figure 2-86. Courtesy, The University of Nebraska.

Figure 2-87. Courtesy, Texas Parks and Wildlife Department.

Figure 2-88. Courtesy, Myrick, Newman, Dahlberg and Partners, Inc.

ISSUE: "SHOULD MY AGENCY DESIGNATE A VOLUNTEER COORDINATOR?"

A central authority with responsibility for the entire volunteer program, such as a volunteer coordinator, greatly improves the program's efficiency. Recruitment, screening, placement, orientation, training, recognition and evaluation are typically the responsibilities of the volunteer coordinator. Volunteers can be assigned on a priority basis and quickly reassigned as tasks are completed or program adjustments are made.

The potential that can be activated through the efforts of a centralized volunteer coordinator is substantial. One need only look at the record of accomplishment in park and recreation departments such as those serving the Hayward, California, area and the cities of Seattle, Washington, and Anaheim and Upland, California, to substantiate the claim.

An alternative organizational setup is for the volunteer coordinator to serve the needs of a number of agencies. Such is the case in San Leandro, California, where the volunteer coordinator recruits for all city agencies. Upland, California, even goes one step further and recruits for appropriate private nonprofit groups as well. Again, the results speak for themselves.

A third approach has been used with notable success in Baltimore County, Maryland. There, decentralized recreation and parks councils have been set up in communities within the county. Although each council acts independently of the others, all volunteer efforts for a given community are coordinated through its own council. The council then functions as the volunteer coordinator for the community.

Figure 2-89. Courtesy, U.S. Department of the Interior.

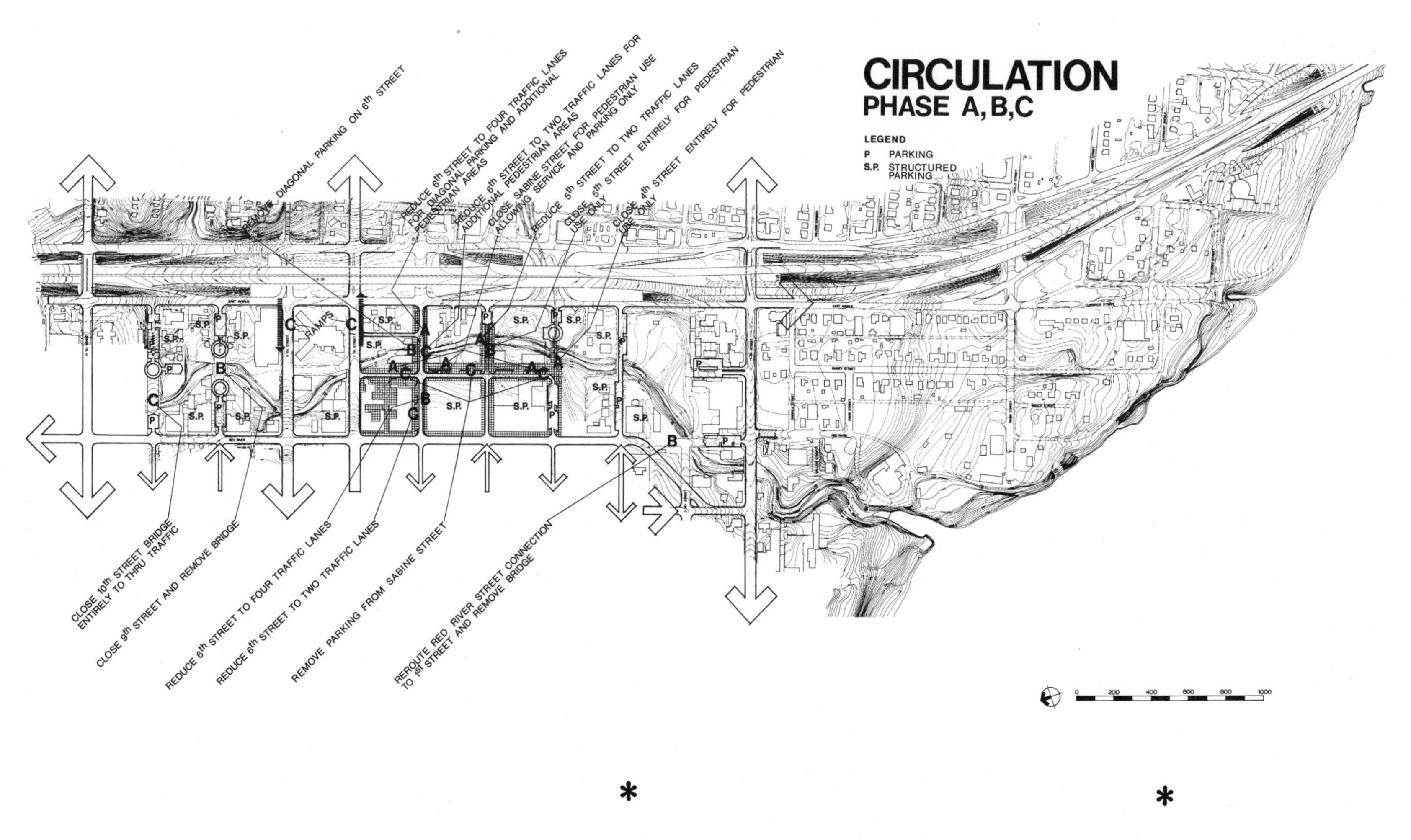

*　　　　　　　　*

Figure 2-90. Courtesy, Myrick, Newman, Dahlberg and Partners, Inc.

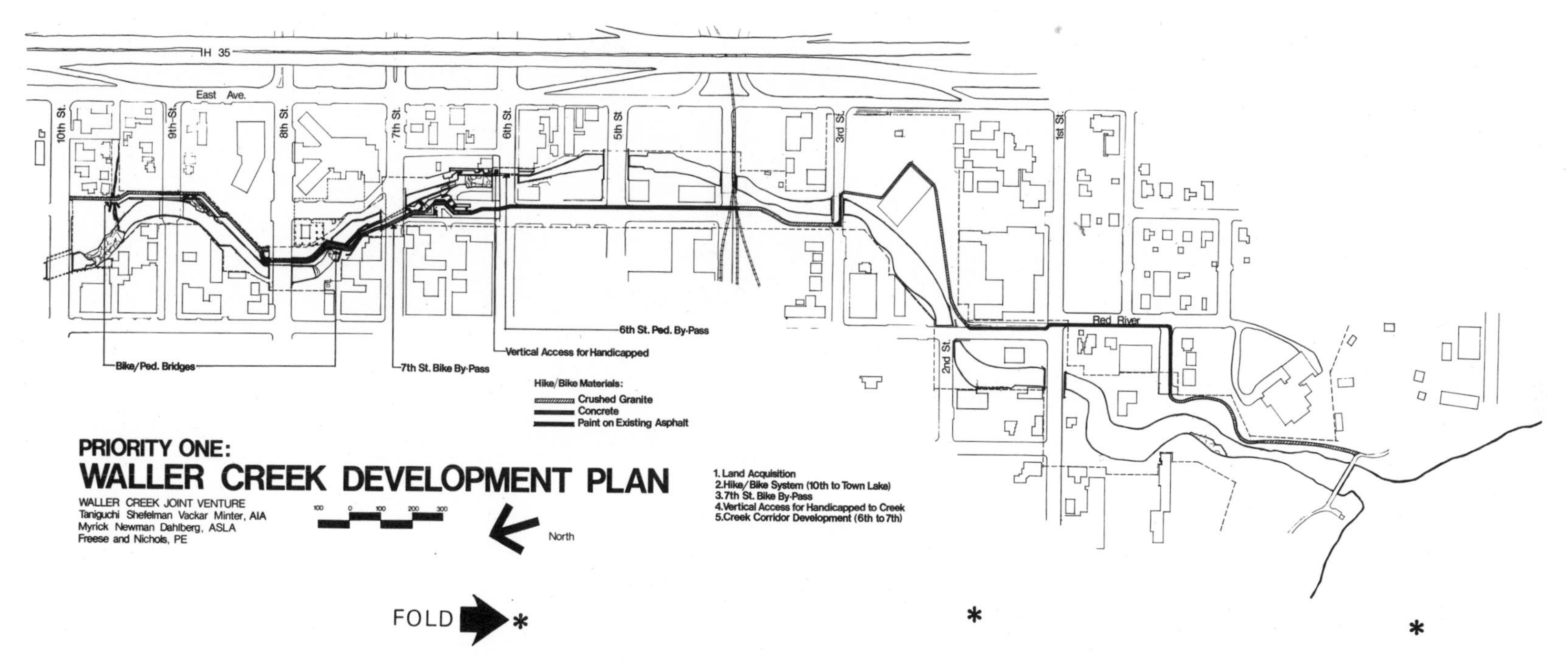

Figure 2-91. Courtesy, Myrick, Newman, Dahlberg and Partners, Inc.

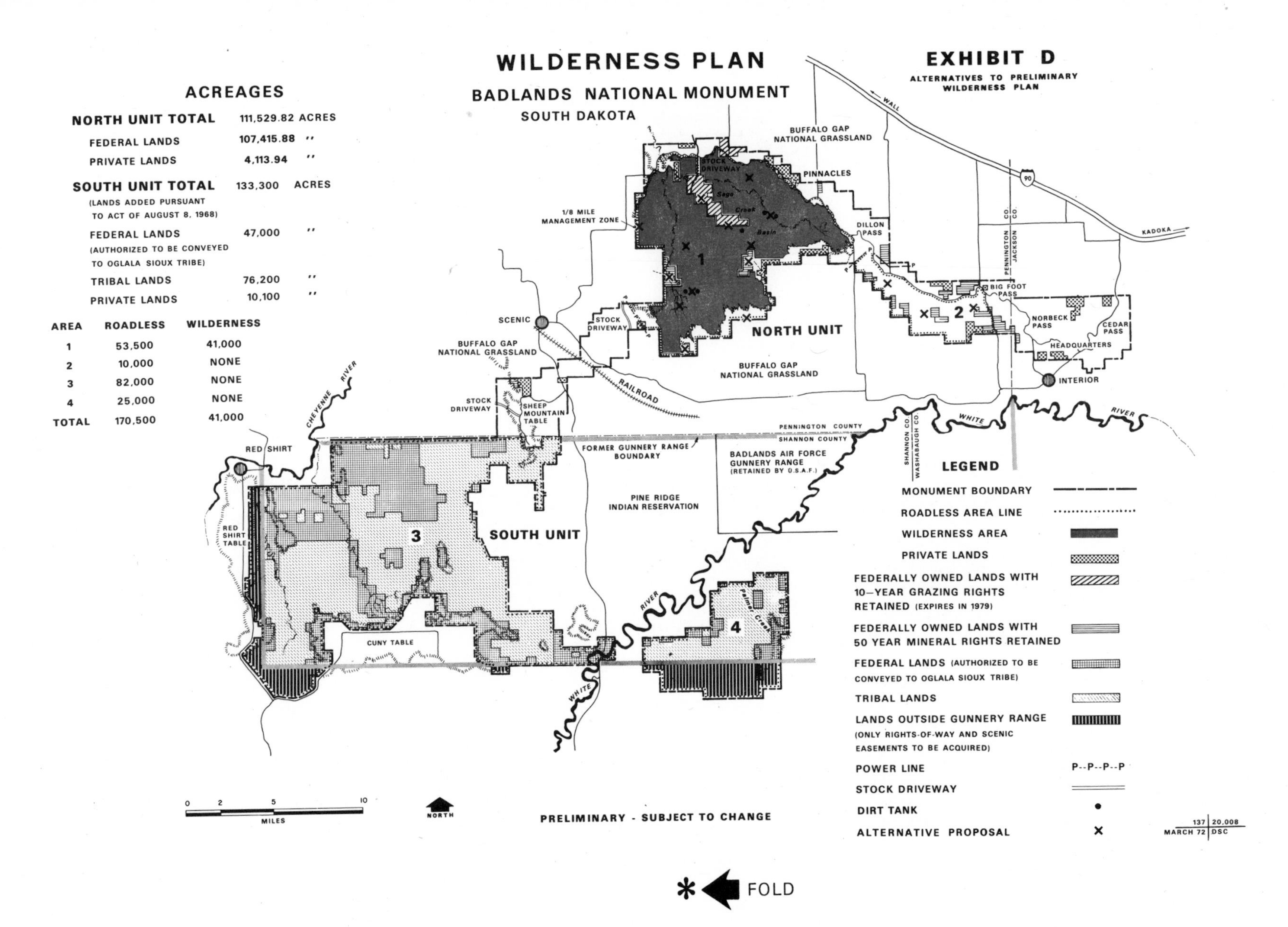

* FOLD

Figure 2-92. Courtesy, U.S. Department of the Interior.

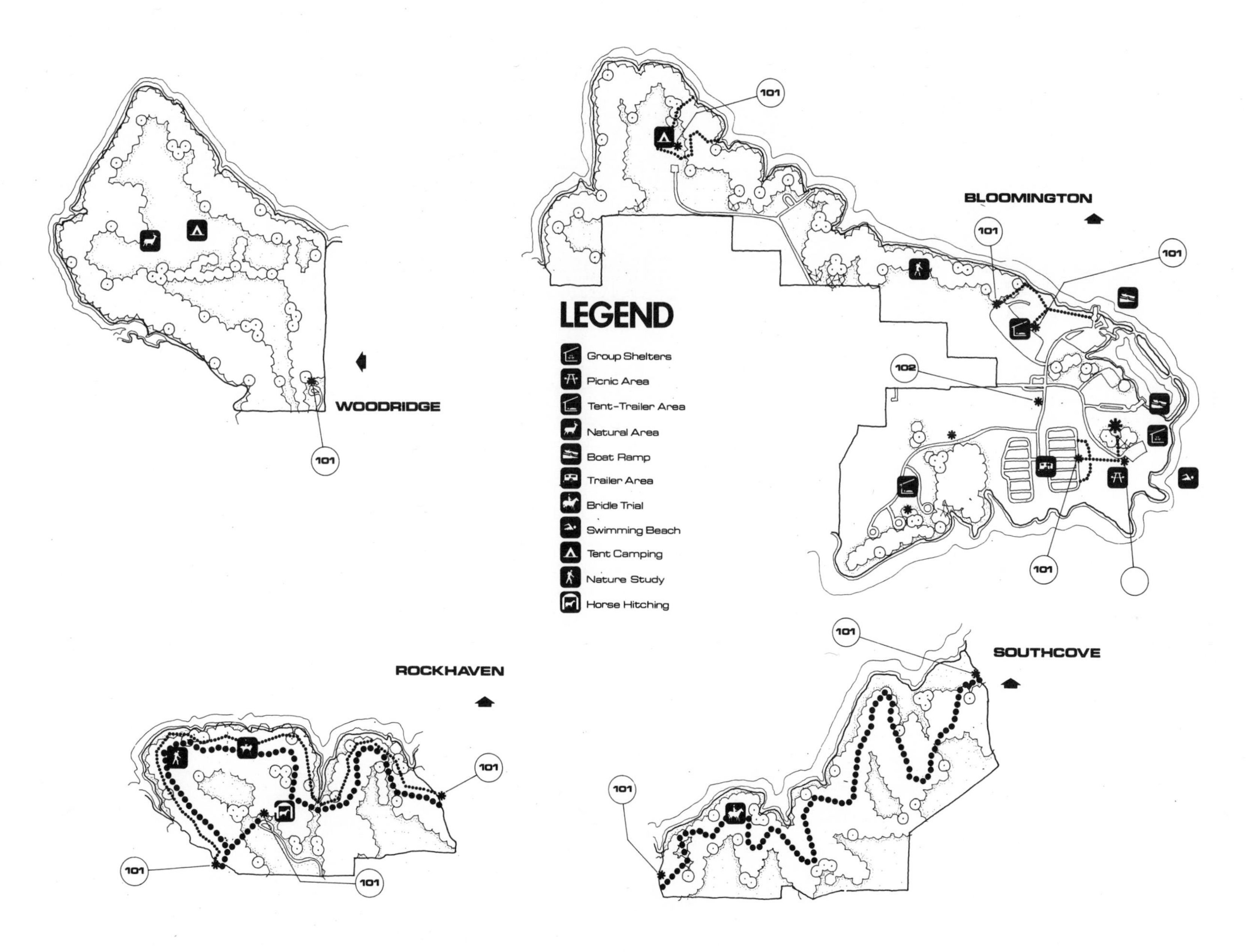

Figure 2-93. Courtesy, Theraplan, Incorporated.

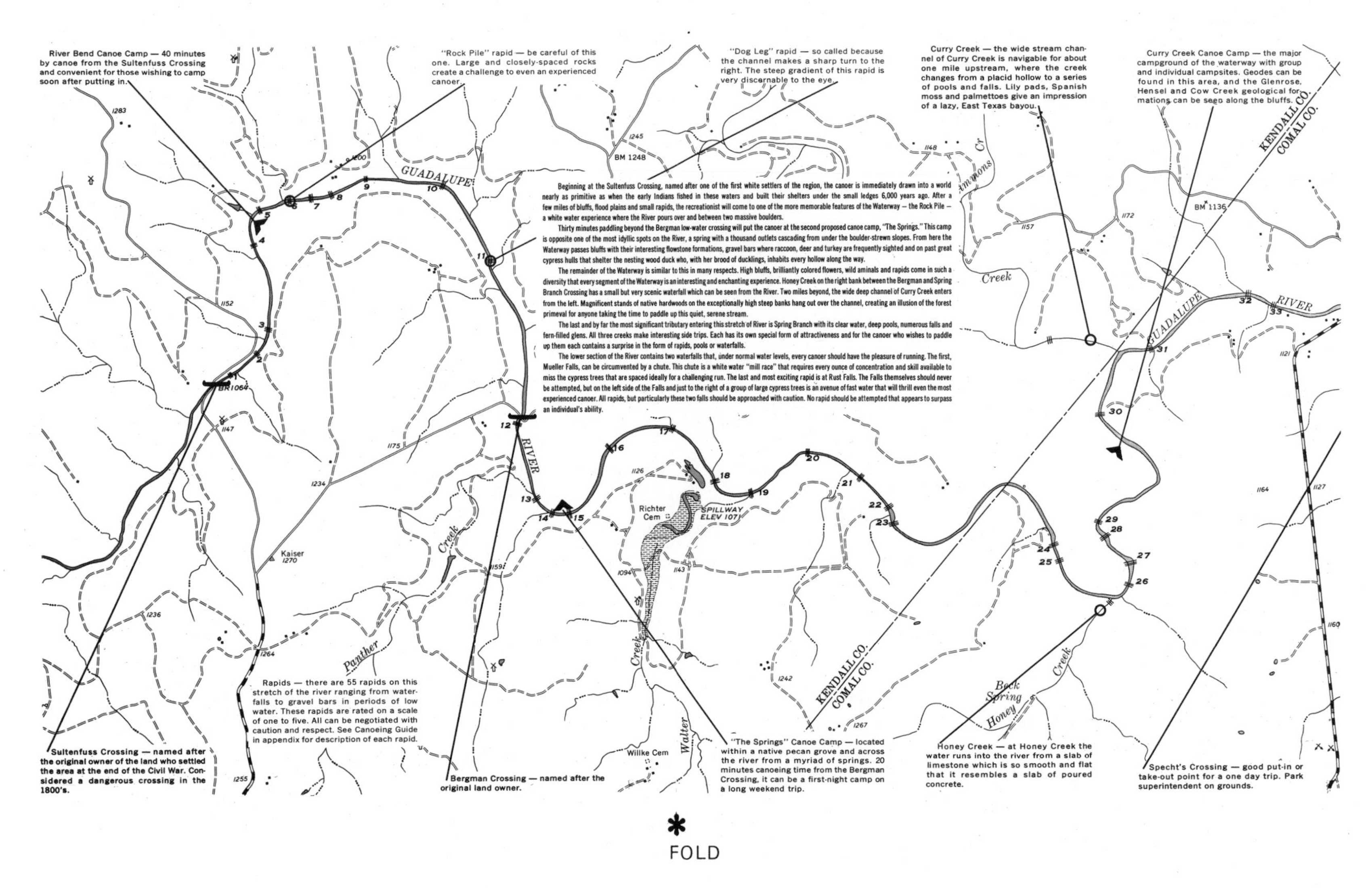

*

FOLD

Figure 2-94. Courtesy, Texas Parks and Wildlife Department.

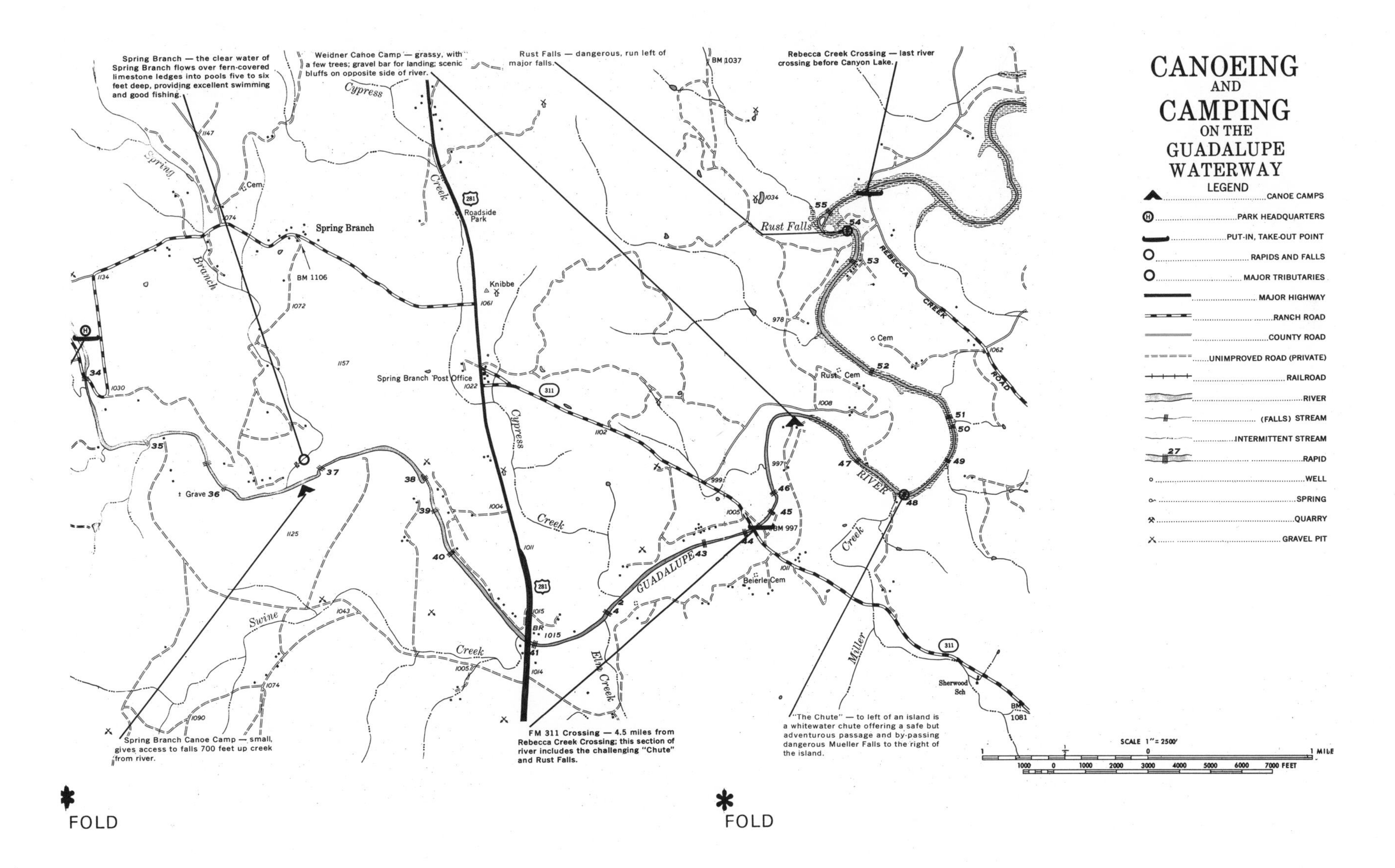

CANOEING
AND
CAMPING
ON THE
GUADALUPE
WATERWAY
LEGEND
CANOE CAMPS
PARK HEADQUARTERS
PUT-IN, TAKE-OUT POINT
RAPIDS AND FALLS
MAJOR TRIBUTARIES
MAJOR HIGHWAY
RANCH ROAD
COUNTY ROAD
UNIMPROVED ROAD (PRIVATE)
RAILROAD
RIVER
(FALLS) STREAM
INTERMITTENT STREAM
RAPID
WELL
SPRING
QUARRY
GRAVEL PIT
Spring Branch — the clear water of Spring Branch flows over fern-covered limestone ledges into pools five to six feet deep, providing excellent swimming and good fishing.
Weidner Canoe Camp — grassy, with a few trees; gravel bar for landing; scenic bluffs on opposite side of river.
Rust Falls — dangerous, run left of major falls.
Rebecca Creek Crossing — last river crossing before Canyon Lake.
Spring Branch Canoe Camp — small, gives access to falls 700 feet up creek from river.
FM 311 Crossing — 4.5 miles from Rebecca Creek Crossing; this section of river includes the challenging "Chute" and Rust Falls.
"The Chute" — to left of an island is a whitewater chute offering a safe but adventurous passage and by-passing dangerous Mueller Falls to the right of the island.
Spring Branch
Spring Branch Post Office
Roadside Park
Knibbe
Cypress Creek
Swine Creek
Elm Creek
Miller Creek
Rebecca Creek Road
Guadalupe River
Rust Falls
Rust Cem
Beierle Cem
Sherwood Sch
Grave
Cem
BM 1037
BM 1106
BM 997
BM 1081
281
311
SCALE 1" = 2500'
1 MILE
1000 0 1000 2000 3000 4000 5000 6000 7000 FEET
FOLD
FOLD

Although a railroad right-of-way is periodically burned, disturbing flora to some degree, it is fenced to protect the area from stock. In the words of Roy Bedicheck in *Adventures With a Texas Naturalist*: "Unconsciously, and certainly with no such purpose in anybody's mind, the fencing-in of automobile and railroad right-of-ways has created far-stretching arboretums without which many species of plants would have been lost, at least to certain localities and even to certain regions, as well as the animal life depending upon them for survival.

"The lanes of pioneer times did not serve this purpose to any great extent, since lands and commons were ordinarily grazed more closely than contiguous pastures. But from the very first, railroads had to fence their right-of-ways to avoid paying one hundred dollars for damages for killing a ten-dollar calf."

From the ledges of the Cibolo Creek Crossing grow two of the rarest shrubs in Texas—*Philadelphus serpyllifolius*, the thyme-leaf mockorange and *Styrax platanifolia*, variety stellata, the hairy sycamore-leaf snow-bell.

Styrax platanifolia—once a common flowering shrub along the bluffs—is now nearly extinct, with only one blooming specimen remaining. It is possible that the plant may be growing in a protected spot on the right-of-way where animals and fire have not been able to destroy it.

Because these two shrubs are very rare they should be protected from people and animals. It is therefore recommended that measures be taken to prevent their collection by amateurs or their destruction by stray stock.

One of the more delightful aspects of travelling through the Edwards Plateau is the roadside plants which are in bloom almost all year round. In spring, fields of blooming Indian blanket and sunflower add such a grandeur to the countryside that one may forget the less spectacular plants blooming around him. In summer the shooting star and the purple nightshade can be found as well as a shy and somewhat rare "red and yellow columbine" which grows from moist crevices of the limestone ledges. Fall brings out the blazing star, snow-on-the mountain and acacia to name several. Only during the short winter season does the landscape lack for flowers.

The corridor or linear aspect of any trail or waterway lends itself well to nature study because one can travel leisurely along them, stopping at will to observe things of interest; in the words of John Steinbeck: "We find after years of struggle that we do not take a trip; a trip takes us. Tour masters, schedules, reservations, brassbound and inevitable, dash themselves to wreckage in the personality of the trip."

30

Figure 2-95. Courtesy, Texas Parks and Wildlife Department.

Figure 2-96. Courtesy, Texas Parks and Wildlife Department.

Committal Service Shelters

The Committal Service Shelters are located adjacent to the Wetland Corridor in a mature oak and hickory grove. This unique area was chosen for its quiet atmosphere, full canopy trees, and outstanding vistas out over the pond and marsh. This attractive setting will leave a lasting impression on families, friends, and other visitors.

The shelter sites are situated off the primary road on a one-way loop drive with a pull-off lane for ten cars at each shelter. Existing native plant material between shelters is retained to ensure privacy. The shelters are oriented for maximum protection from inclement weather as well as taking full advantage of surrounding views.

The shelter's simple shed roof and warm tone materials are in character with the cemetery's architectural theme established in the Administrative Building and in harmony with the surrounding environment.

24

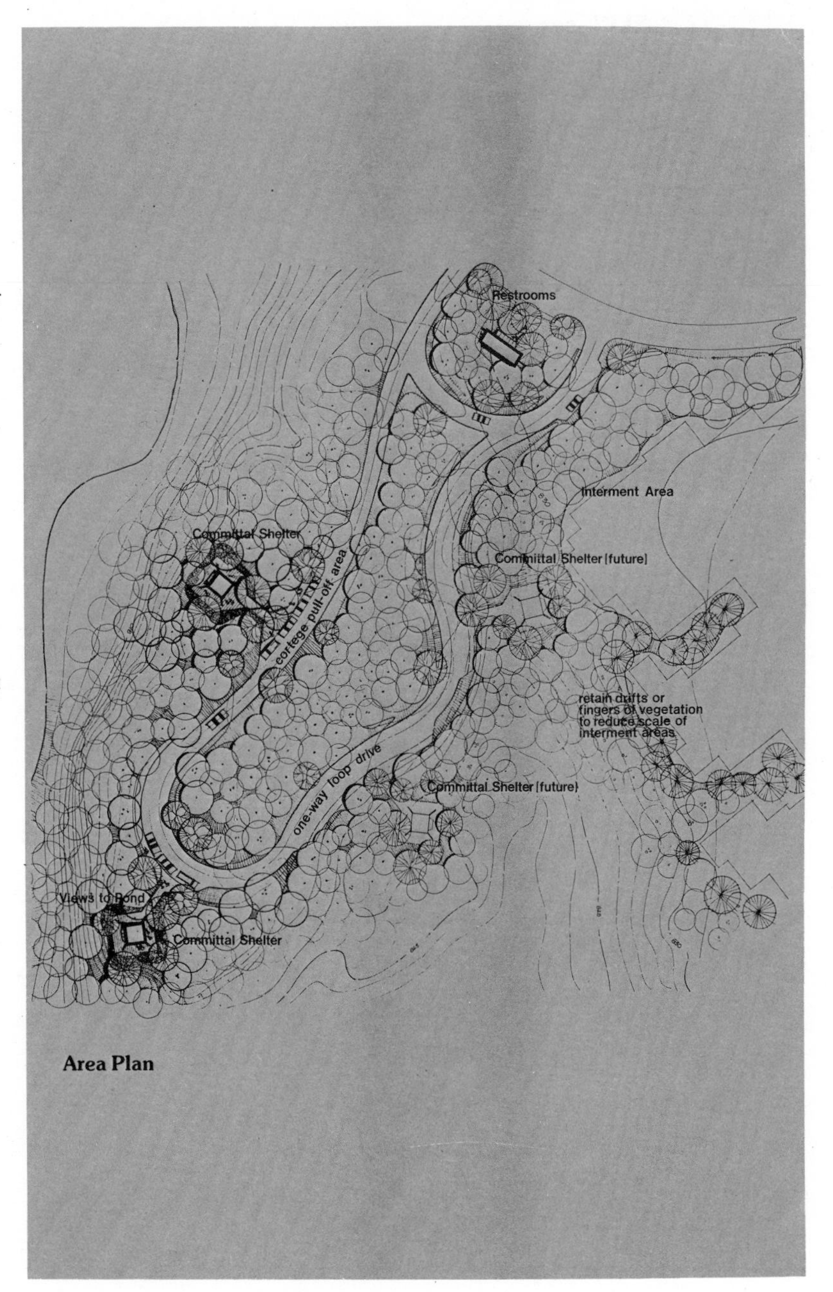

Figure 2-97. Courtesy, Johnson, Johnson and Roy, Inc.

DESIGN GUIDELINES

Waller Creek is almost lost in the confusion of existing sub-par facilities, sewer channelization, deteriorating corridor environs and an abundant road network. Initial development efforts will focus on improving the use of the creek area and its image. By providing proper access and visibility, the community will become aware of its location and the available recreational opportunities. Within a proper design framework, repetition and consistence in details can provide a sense of continuity and orderliness. To this end, specific design guidelines have been developed to be considered as detail designs are developed for each area. The guidelines were developed in consideration of the creek ecology, the physics of water movement, safety to the user, aesthetics and spatial feeling, development character and integrity, and many other factors which contribute to overall exterior design. The character of existing development has been maintained and enhanced where it contributes to the creek environment.

Guidelines for the entire creek corridor have been developed to provide the desired continuity of detail. They relate to the many facets of the landscape design palate, including:

- Plant materials
- Grading design
- Paving
- Structures
- Landscape furniture
- Lighting
- Graphics
- Water features
- Accessories

LANDSCAPE ZONES

The above elements of design fall into four basic categories or zones that define the landscape character. These zones have been established based on the level of natural and man-made improvements to the creek and have been designated as:

Urban
Hard
Soft
Open

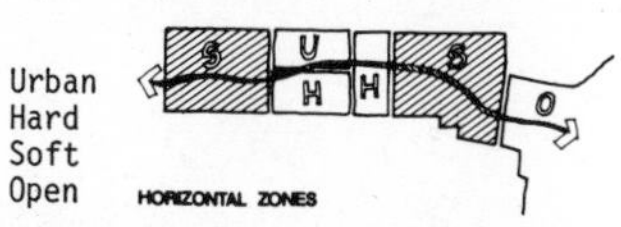

These zones are used to define and establish the quality of all future development throughout the corridor. Design and planning responses should respect these zones, as they are vital to the development of a strong visual unity. It is important that the details and the standard vocabulary be consistent, but not inflexible where the needs of a special area require a more unique solution.

A special zone category has been included to detail those elements which are designed to be continuous throughout the length of the corridor.

Urban Landscape Zone

The existing development and the man-made creek edges imply a tight spatial feeling. This and the upper grid level pattern of the streets represent a more "urban" feeling. The general design attitude should reflect the density of the development and consequent activity by broadening the creek edges and adjacent walks and plazas. Vertical access to the creek itself is extremely important and therefore within the more restrictive creek edge, limitations require a much more structured design response.

Hard Landscape Zones

In this zone, the concern is transition of the development as it moves to the less dense areas. The structural elements necessary to reach the creek edges are crisp or "hard".

51

Figure 2-98. Courtesy, Myrick, Newman, Dahlberg and Partners, Inc.

One of the most significant changes in local government in recent years has been a growing awareness of the need for interlocal cooperation. Cities, towns, counties and school districts that serve a common area are joining together in regional efforts to solve mutual problems. This move to regionalism is due to three basic reasons:

1. The nature of the community has changed. No longer is it restricted to the boundaries of a single city or county, whether rural or urban. The rapid growth and modernization of our nation has brought with it a highly mobile, highly demanding public which is involved daily with a number of local governments. Today's citizen rarely works, lives, shops and enjoys his recreation within a single jurisdiction.
2. The increasing cost of providing governmental services is requiring that local officials pool administrative operations on a broader geographic basis for economy and efficiency.
3. Many problems facing local governments do not adhere to political boundaries and are thus multijurisdictional in scope. These problems include air and water pollution, transportation, economic development, law enforcement, health protection, solid waste disposal and others. Effective solution to such problems necessarily requires coordination and cooperation between those jurisdictions affected.

One of the most promising mechanisms for regional cooperation is a Council of Governments (COG). While the COG concept is not entirely new to Nebraska, it appears that, in general, understanding among public officials and citizens is lacking. The purpose of this introductory brochure is to answer some of the more commonly asked questions concerning COGs.

1

Figure 2-99. Courtesy, The University of Nebraska.

Figure 2-100. Courtesy, Johnson, Johnson and Roy, Inc.

to the assembly and rostrum area from the entrance drive. A problematic, badly eroded section of the stream will be eliminated by enclosure and filling. The proposed enclosure structure would provide an opportunity to reduce the velocity of the stream by installing concrete energy dissipators in the bed of the stream. These structures can be designed so that water is not retained during low flow. The structure recommended for the enclosure is an elongated corregated metal arch fastened to a concrete foundation on both sides of the present stream bed. Except for minor realignment and the installation of concrete dissipators, the existing stream bed would be used as the flow line invert of the enclosure. This plan for enclosure will eliminate the existing stream habitat for 400 feet.

Implementation of the proposed development plan will result in an increase in runoff water from the site due to establishment of paved surfaces and thinning of the leafy canopy. Erosion problems are not anticipated since grass will protect soil on developed upland sites, and undisturbed forest floor vegetation and leaf litter will continue to protect soil in the sinkhole ravines.

Storm drainage from roadways will be collected in storm sewers. The roadway cross section will be an inverted crown, sloped transversely to carry storm water in the center of the roads. This arrangement will prevent salt which is applied for de-icing from draining onto vegetated areas. Full length trench drains constructed at intervals will intercept storm water flows and divert water into connecting storm sewers located in the center of the roads. The trench drains will be constructed wide enough to permit occasional cleaning of the system. Because of the steep gradient of most of the roadways and the presence of highly erodible soils, a stilling well is proposed for all discharge points into the drainage channels. To reduce the amount of storm water flow into sinkholes that are to remain in in their natural state, storm sewers will be extended in many locations, past the natural drainage ridge line. The construction of relatively deep manholes is required to transfer the drainage by gravity from the sinkhole watersheds to the storm sewer system.

Any possible reduction in ground water recharge which might result from development of the cemetery will be offset by porosity of pool bottoms. Even though pools are designed to hold water, some leakage is inevitable. Since excess storm water will be discharged to the creek, it will be only partially retained in pools, and much of it will recharge locally.

Bridges are designed so that they will not interfer with natural drainage patterns or destroy the delicate banks of sinkhole ravines. Aesthetically appropriate natural stone foundations will be constructed well away from stream beds. Steel beams will span the ravines so that no filling will be necessary. Erosion will have to be carefully controlled during construction of abutments.

[The effect of burials on ground water quality should be investigated and discussed in an Environmental Impact Statement.]

Flora and Fauna

The value of the undeveloped Jefferson Barracks Cemetery property to native Missouri flora and fauna will be diminished by implementation of the proposed plan. The diversity of habitat types will be reduced by development,

15

Figure 2-101. Courtesy, The University of Nebraska.

5) **Restore cornice work, window trim, decorative lintels, etc.** to enhance the rhythmic quality of the downtown area. Paint and revitalize those buildings that have not yet received this care. These details bring out the character and uniqueness of Council Grove.

6) **Continue the rhythmic elements around the buildings on the corners of Main Street, to the adjacent sidestreets.** Similar roof lines, details, openings for windows, etc. are just as important for continuity on the side streets. These reoccuring elements, along with compatible materials can help unify the core and distinguish the downtown area from surrounding areas.

7) **The architectural details should not be covered, hidden, or altered in any way** if the present character of the community is to remain intact. These fine building details should be seen and enjoyed by all. False fronts hide the original architecture of a building, and often detract from neighboring original facades. However, if false fronts are desired, use complementary materials and color schemes that will not result in the disruption of the architectural rhythm and character of the area.

8) **Integrate sign heights, color schemes, and materials used** to lessen the affect that signs have upon the rhythm of the downtown area. Signs that extend perpendicularly from the building facade should be carefully chosen, as they are the most visible. As the purpose of signs is to attract the eye, they can easily disrupt the overall rhythm of an area. However, if carefully handled, they can serve their function and at the same time enhance the architectural character of a commercial district.

9) **Give careful consideration to new buildings or other additions to the downtown area.** Use complementary materials (stone, brick, concrete, etc.) in new construction. Try to enforce the positive use of the rhythmic elements of Council Grove—roof lines, building heights, opening sizes, etc.

10) **Attract new and varied uses of the downtown area in order to assure its economical productivity in the future.** If possible, explore new ideas and uses for the vacant second floor spaces, presently used primarily as storage areas. Many contemporary uses can easily be adapted to older structures that have outlived their original uses.

11) **Explore the possibility of expanding the present historic trail to include the downtown area,** which is a valuable example of turn-of-the-century architecture. The physical history, historical growth and resources of Council Grove should be emphasized and publicized along with the other designated landmarks in the community.

Individual building elements form vivid rhythm on Main Street

CONCLUSIONS

47

Figure 2-102. Courtesy, Kansas State University.

At the same time several one and two story low-rent housing structures for non-elderly families were built with federal assistance in the south part of town, east of First Street between Park and Race Streets. The ten units are considered adequate for the community's present needs.

Library

A modern, all-masonry constructed library that provides adequate service to the community is located on the northwest corner of Fourth Street and Pear Avenue.

Post Office

Deshler has a well cared for Post Office building that was built in 1950 and has been leased by the United States Post Office Department since that time. This structure is centrally located on Fourth Street between Railway and Bryson Avenues.

The current postal staff includes 2 full-time employees (including the postmaster) and 3 part-time employees.

Medical Facilities

County hospital facilities are provided at the county seat, Hebron. Deshler is in need of a dentist now, and the doctor is close to retirement.

Nursing Home

A municipally owned 51-bed Nursing Home, opened in 1968, is now operating at full capacity under the guidance of the Good Samaritan Society. This facility is located on South Fourth Street between Park and Race Streets.

Parks and Recreation

A Public Park located north of Thayer County Fairgrounds in the southeast quartile of the town contains a swimming pool, playground equipment, picnic tables, fire pits and shelters. The grounds and equipment are well maintained. Washington Park is a small older facility that is located just south of the business district on Fourth Street. In this park is situated the commemorative marker for the founder of Deshler. In the planning section of this report, an expansion of the present park falilities will be proposed.

Churches

There are three churches in the community. St. Peter's Lutheran Church, a modern facility dedicated in 1963 is located on the northeast corner of Hebron Avenue and Fourth Street. Peace Lutheran Church also has a new building that is lo-

16

Figure 2-103. Courtesy, The University of Nebraska.

Figure 2-104. Courtesy, Johnson, Johnson and Roy, Inc.

with occasional clumps of Rocky Mountain juniper. Wildlife includes mule and white-tail deer, antelope, the Badlands chipmunk, the black-tailed prairie dog, the Western meadowlark, the lark bunting, the Western painted turtle, and the reintroduced buffalo and Rocky Mountain bighorn sheep. Erosion continually exposes evidence of prehistoric environments in the form of fossils belonging to the Oligocene epoch and Cretaceous period. The North Unit is surrounded in all directions except to the northwest by the Buffalo Gap National Grassland. These lands are managed principally for grazing by the U.S. Department of Agriculture-Forest Service.

Local ranchers lease grazing rights on the national grasslands, supplemented by their own grazing lands and croplands, thus creating a patchwork pattern of landownership around the monument. Grazing is also the principal land use in the South Unit, on both Oglala Sioux Indian tribal lands and Federal units (which include those under life estate provisions). Both tribal and individual lands suitable for grazing are leased to Indian and non-Indian cattle operators under provisions established by the Oglala Sioux Tribe and the Bureau of Indian Affairs.

Scenic and recreational facilities are grouped in three general regions around the monument: the Black Hills region includes one large national forest, four National Park System areas, two State parks, fish and game recreation areas, and numerous private campgrounds and trailer parks. Along the Missouri River to the northeast, in central South Dakota, are three large river reservoirs that provide excellent water-oriented recreation. Badlands National Monument and the Pine Ridge Indian Reservation together comprise a third recreational region, with potential that is only beginning to be realized.

16

Main access to all three regions is by Interstate 90, a major east-west route to Yellowstone and Grand Teton National Parks from the upper midwestern United States.

One role Badlands National Monument plays in its region is to introduce easterners to a representative unit of the National Park System. The monument is the threshold to the Indian reservations and the Black Hills recreational complex.

About 1,330,000 people visit Badlands National Monument annually.

This addition lies to the south and west of Badlands National Monument as it existed prior to August 1968, and is now designated

Figure 2-105. Courtesy, U.S. Department of the Interior.

Figure 2-106. Courtesy, Johnson, Johnson and Roy, Inc.

Organization, commitment, and planned strategy are essential, even in the smallest projects.

The second step was to identify resources. Potential donors contacted were the local businessmen in the area. In this case, solicitations were made by phone, and, within hours, the project goals were reached. The success of this "against-the-rules" approach is based on professional and successful selling. The case clearly presented the needs of small children, always a good cause, and the requests were specific and appropriately matched to the donor who could supply them. For instance, a large lumber company based in Kansas City, Missouri, with branches in Colorado offered to supply and deliver lumber. A tire company offered to provide tires for various structures and to provide transportation of the goods to the site. Other local businesses made cash contributions. A very creative and different approach presented itself when a local bank was solicited. The bank donated funds and then challenged two local realty firms to match that donation. The realty companies responded to the challenge. In return for their generous contributions, the businesses will receive tax benefits, local publicity, and recognition of their involvement in community affairs.

The committee will have an integral part in the construction of the park, and volunteer services of the children's parents will be tapped. The Park and Recreation Department will donate landscape architectural services. Donating sponsors will be properly thanked at a formal dedication of the facility, complete with news coverage by the local media. The preschool playground is just one example that private sector fundraising techniques work by using planning, organization, coordination, and effort. With a well-organized and presented project and a little perseverence, one can ask for help and expect to get it.

EAST OAKLAND YOUTH DEVELOPMENT CENTER (Corporate Initiative)

East Oakland is a growing urban area east of the San Francisco Bay region. Its Youth Development Center is an excellent example of how a major corporation can set an example for developing needed community/public services

The idea originated with top management in the Clorox Company who wanted to do something in the area that was the first home of Clorox. The company chose to focus on the needs of young people between the ages of 12 and 18, a group which they felt needed attention and could use better facilities.

Groundwork for the project started when Clorox donated land and hired architects to develop the plan. The Center will be 20,000 square feet and is designed to accommodate about 500 teenagers a week.

From the inception of the Youth Development Center project, Clorox wanted to demonstrate that businesses and foundations could work effectively together to supplement government programs aimed at meeting social needs in the community. Clorox realized that only a joint effort could succeed and that participation by other contributors and the residents themselves was essential.

Phase I of the fundraising was to raise $1.5 million for construction. Clorox's fundraising campaign focused on three major approaches.

27

Figure 2-107. Courtesy, U.S. Department of the Interior.

Treatment of Edges

Alternatives for edge treatments represent conditions achieved either by clearing existing vegetation, introducing additional plant material or by a combination of these strategies. Visitors will enjoy the contrast and variety of mass, color and texture of woodland edges.

Low Shrub Understory

Damaged or undesirable plant material is removed and replanted with shrubs which will ultimately reach 20″-40″ under the existing higher canopy. This permits unobstructed views while prohibiting physical access.

Full Understory

A sense of enclosure is achieved by retaining or establishing a full understory of shrubs to the edge of a tree mass.

Meadow Grass Understory

By removing understory shrubs and undesirable canopy trees and replanting with meadow grass under the canopy trees, visual space is extended, offering quite a different feeling from the other edge treatments.

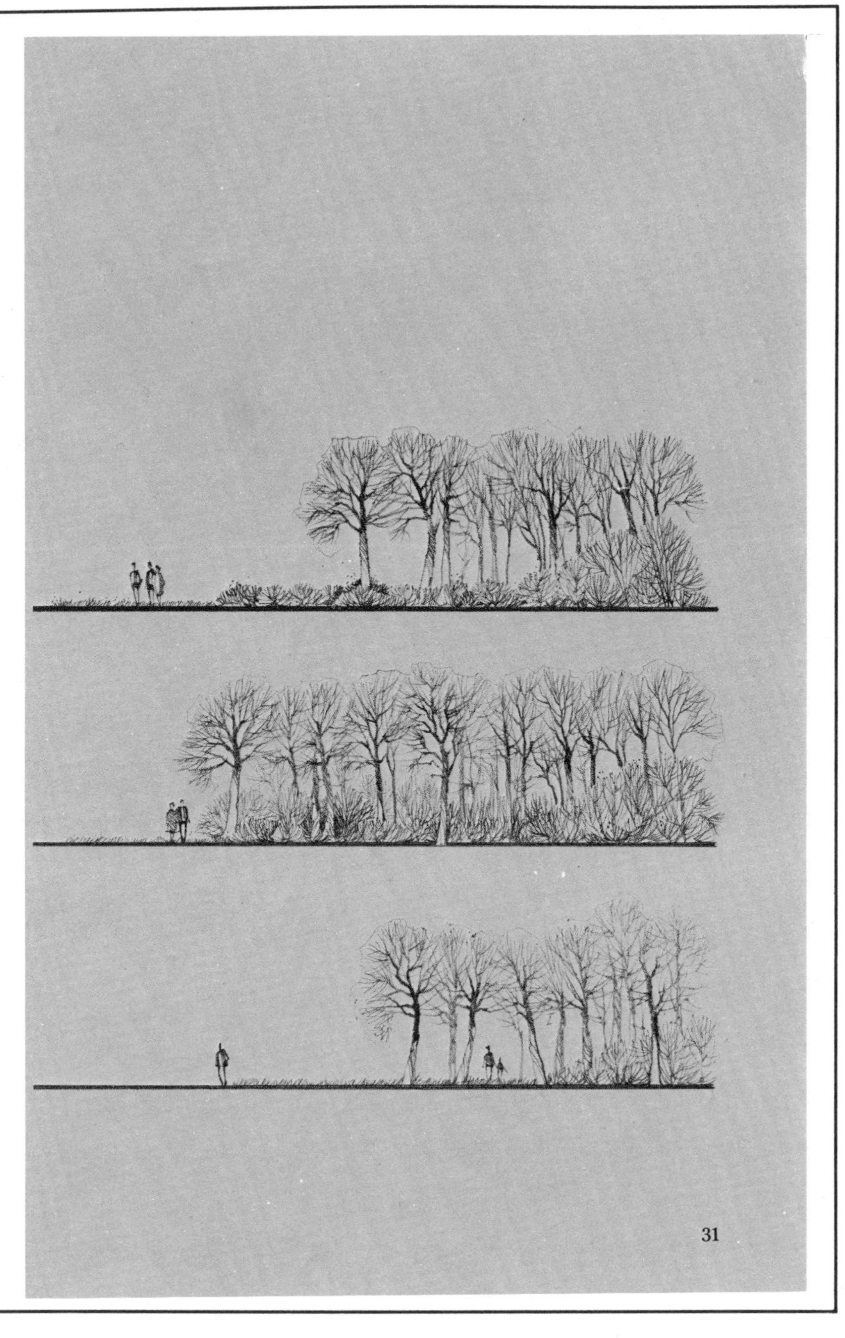

Figure 2-108. Courtesy, Johnson, Johnson and Roy, Inc.

4. Distribution

The distribution of the catalog should be to as many people in the community as possible. If prime potential contributors have been identified as part of a systematic private sector involvement program, a special mailing might be made to this key group first. If possible, of course, a donation for the distribution should be secured. In nearly every city, there are firms which specialize in mass mailings, and they should be approached.

The Mt. Holly Parks and Recreation Committee prepared a number of different cover letters to personalize the appeal of gifts catalogs sent to the various segments of its community. The business community received a letter speaking in terms of hard economic realities and highlighting the public relations and income tax deduction advantages of giving through the gifts catalog. Civic organizations received a letter oriented to their service ethic and almost without exception responded with a cash gift to the township. Individual residents, identified from voter registration lists, were sent an appeal based on their desire to "make a difference" in the quality of recreation in Mt. Holly. The $5 - 10 donations continue to arrive, nearly a year after the catalog was distributed.

Possibly the gifts catalog could be mailed to households in the community as an enclosure with water bills, bank statements, credit card billings, retail catalogs, or even with property tax assessments. Maybe the local newspaper would include the catalog as a supplement to a special edition. If your agency or organization must cover the cost of distribution of the gifts catalog, there are ways to defray the normally high cost of a mass mailing campaign. Bulk rates, available through the Post Office, can substantially lower the mailing costs.

The catalog can also be made available in libraries, at schools, in banks, at retail business establishments, an on buses. A special effort should be made to provide gifts catalogs to attorneys and trust officers who frequently deal with people designing legacies. What better legacy than a gift to benefit public recreation or historic preservation?

8

Figure 2-109. Courtesy, U.S. Department of the Interior.

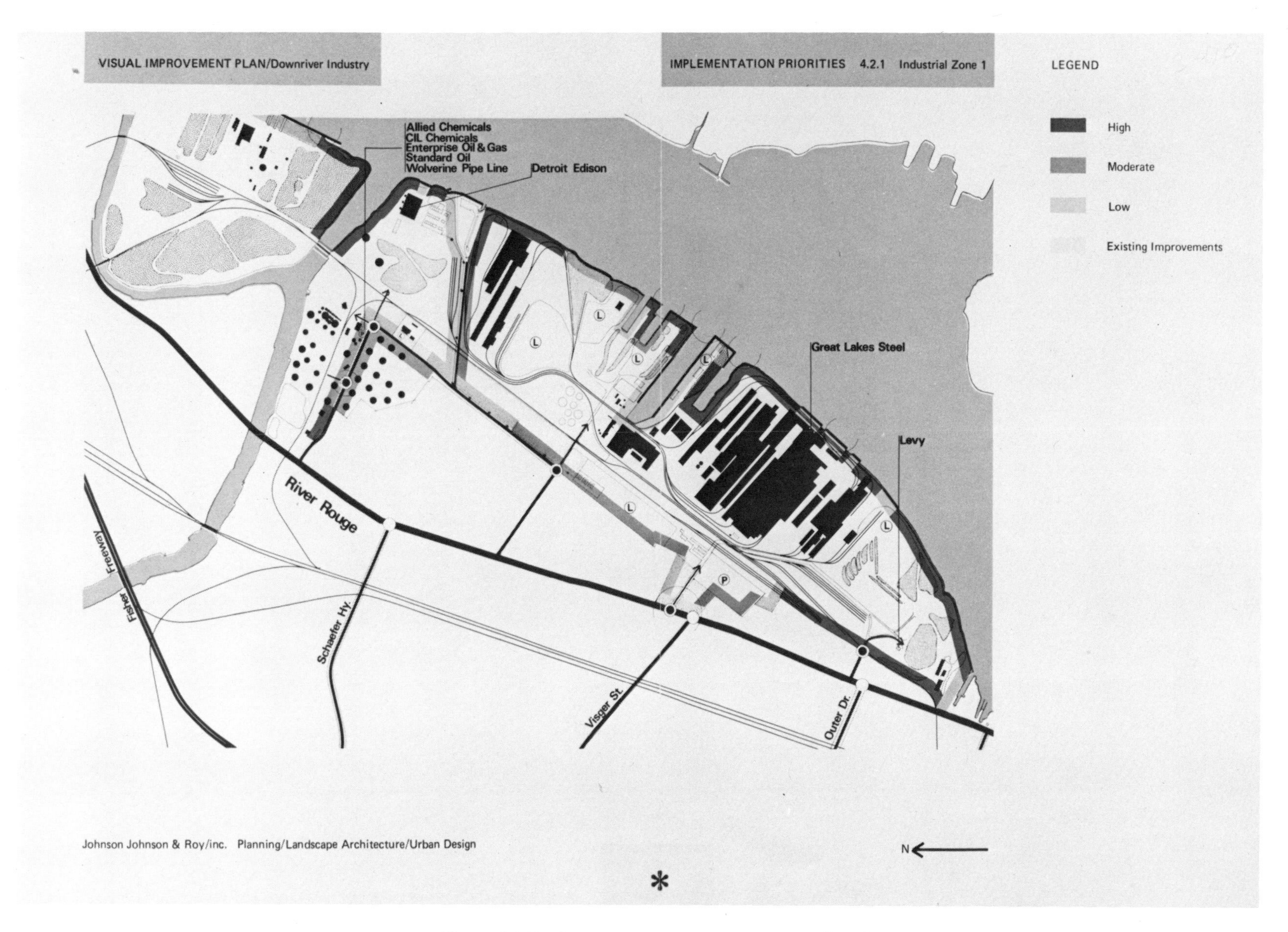

Figure 2-110. Courtesy, Johnson, Johnson and Roy, Inc.

The Long-Term Solution

The humanities and performance complex we envisage will consist of:

A Center for Arts and Humanities

An adjoining University Performance Hall

A **Center for Arts and Humanities** will occupy the completely renovated athletics building on the Michigan Tech campus. It will accommodate programs which combine teaching and research in communications with the study of literature, foreign languages, and the philosophy of science. It will link the humanities to programs in both the arts and technology, accommodate staff in art and design, and provide integration of technical communication, production, and performance design.

A *Theater for Arts and Technology* will also occupy this building, providing a performance space designed to complement the technical mission of the University. It will have a flexible "production space" including a configuration of stage and seats adaptable to the needs of each work. Such space will encourage students, faculty, and artists to explore the interrelationship and adaptations of technology to theater art.

The **University Performance Hall** will adjoin these facilities and support the activities of the music faculty, students, and the entire campus community. It will bring to a University-community audience drama, music, and lyric theater for which there presently are no facilities. This hall will be designed to support University-community organizations such as the Keewanaw Symphony Orchestra and will accommodate major visiting performances.

In summary, the building program will provide: 1) office, classroom, and laboratory space for faculty in English, foreign languages, philosophy, and speech in the renovated athletics building; 2) performing, rehearsal, studio office, storage, and gallery space for the faculty in art, music and theater also in the existing structure; and 3) a Performance Hall, with related service areas, in a new addition.

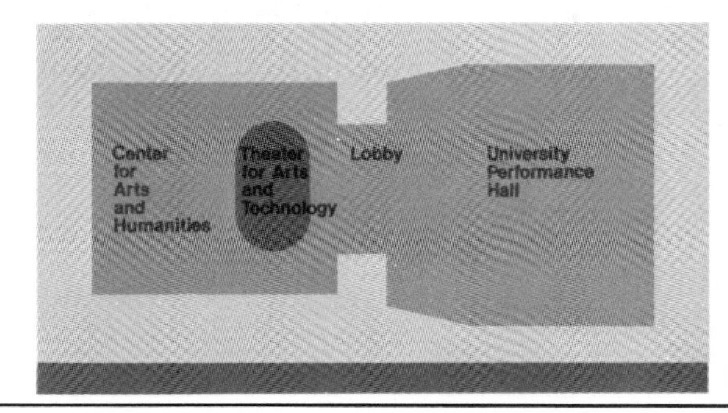

An Ongoing Commitment

Michigan Tech's fundamental commitment is to the continuation and advancement of high quality, rigorous programs of undergraduate education and graduate research to meet the needs of society and to promote the welfare of the state and nation.

Secure in its reputation for excellence in the sciences and engineering, the University has directed resources in the past decade to the Department of Humanities to develop a highly competent faculty of over 50 persons. This faculty has developed degree programs in liberal arts, scientific and technical communications and technical theater. With substantial private support the humanities faculty has created a nationally recognized "writing-across-the-curriculum" program especially designed to improve the communication abilities of science and engineering students. The arts faculty has developed curricular and extracurricular programs in theater, music and fine arts involving large numbers of students, faculty, and community members. Continuing University support and the modern facility we seek will promote even greater development of these programs.

Who Will Benefit

Michigan Tech, through its graduates, serves the state and nation. With improved communication skills and a greater understanding of the role of their professions, these graduates will be better prepared to serve society.

The University serves as the cultural focus for the Western Upper Peninsula of Michigan. The improvements proposed here will provide greater opportunities for the people of the region to attend and to participate in programs in theater, music and the arts. This will enable the University to meet its responsibility for the enrichment of the community.

Figure 2-111. Courtesy, Johnson, Johnson and Roy, Inc.

Goals for Arts and Humanities

The future for Michigan Tech has been charted in a series of Master Plan studies, culminating in 1978 in a Long Range Planning Report. Specific objectives in that plan included developing a university-wide communication skills program, contributing to broadening the education of all MTU students; and developing arts programs and facilities to serve as the cultural, artistic, and performance center for the campus and the community.

Both the enrichment of present programs and the development of communications skills are specifically addressed in the Long Range Plan:

"The major direction for Michigan Technological University in the period to 1985 will be the enrichment and enhancement of existing programs rather than the development of new programs...a major effort also will be made in the communication skills area and in an understanding of the impact of technology on society."

We have had considerable success in meeting these goals for faculty and for programs in communications, the humanities, and the arts.

Our next goal is to develop a facility to support the continuation of these efforts.

The Current Program

The MTU program for the arts and humanities is varied. Program emphasis is on providing a humanistic and artistic component to the education of engineers, scientists, foresters, and business people. MTU is committed to providing these students with a broadened education in the humanities as well as opportunities to participate and perform in the arts.

Michigan Tech is a national leader in communication skills education. The goal of this program is to make every student an effective communicator in personal and professional relationships. MTU's commitment to produce articulate graduates is interrelated with the humanistic concern that all students study the social and ethical issues central to our technological society.

Michigan Tech science and engineering students, along with faculty and community members actively participate in the arts, including a symphony orchestra, festival chorus, three jazz bands, a varsity singers jazz choir, a wind ensemble, pep bands, and four or five major theatrical productions a year ranging from Shakespeare and opera to modern drama.

The Current Facility

In 1981, two-thirds of the faculty in the humanities and arts were moved from scattered locations into a building formerly used for athletics and now designated as the Arts and Humanities Center. The facility is inadequate, but it is the best that can be done with existing University resources.

The humanities faculty, with a national reputation in the teaching of communication skills, currently occupies temporary office space that offers little privacy. The music faculty conducts rehearsals in areas adjacent to faculty offices with no acoustical barriers, and the theater faculty stages its shows in music rehearsal rooms.

The program we offer is excellent; the faculty is active and vital. We need help to develop a facility worthy of both. The program outlined here will accomplish that end.

Figure 2-111 (Continued).

What We Seek

Michigan Tech has completed short-term, preliminary modifications to the athletics building to make it minimally habitable for the Department of Humanities. We seek support to:

1. Totally renovate the athletics building to provide modern facilities for programs in communication, literature, philosophy, foreign languages, theater, and the arts.
2. Construct a performance hall addition to meet the needs of the music program and to provide a facility for visiting orchestras, dance groups, lecturers and other cultural events.
3. Landscape the area adjacent to the arts and humanities facility and provide parking space for approximately 500 vehicles.

Private Support is Necessary

The Arts and Humanities Center will enhance the lives of Michigan Tech students for years to come. University faculty and staff and their families will benefit as will the citizens of Michigan's Western Upper Peninsula.

Concerned individuals, corporations and foundations are invited to participate in making the Center possible through tax-deductible gifts to the Michigan Tech Fund.

Facilities We Will Feature

The Arts and Humanities Center will provide:

- Laboratory facilities for students working in printed media
- Communications and writing laboratories for research and teaching in communications
- Joint seminar rooms for the integration of faculties in humanities and the arts
- Offices for a research and teaching faculty
- Communications laboratories for language skills, foreign languages, and text editing
- Technical theater classrooms and performance theater classrooms
- Experimental laboratory theater, suited to alternative stage and audience layouts with multimedia flexibility
- Theatrical costume and set shops for experimental and performance design
- Rehearsal rooms for a student body active in music performance both as individuals and as members of groups
- Performance hall

Figure 2-111 (Continued).

3. The A/V Report

The use of audiovisual materials by consulting organizations for the communication of design concepts has grown steadily in recent years. Both client and designer are finding this medium more convenient and informative than lengthy printed documents. Attractive photographic techniques will often have a more effective impact on an audience and stimulate program actions more rapidly than will written communications: one picture says a thousand words.

Popular forms of A/V reports include the 35-mm slide presentation, the design exhibit, and the motion picture. Each has its advantages and problems in planning and production, but each will have a different impact upon a target audience.

As in the printed composition, the A/V report should have a defined direction. It must be programmed before production begins in order to obtain an organizational objective. Most are horizontal in nature and are created to sell specific concepts or ideas to clients. The content must present facts, review techniques, or recommend actions.

The audience of the A/V report is usually *general*. Technical or management audiences can be reached just as effectively with a written document as with the A/V presentation. This is not to say, however, that a designer should eliminate these readerships from consideration.

ORGANIZATION

As with the written composition, the A/V report must be well outlined before production begins. From the completed outline, basic categories of information may be arranged in a proper sequence for greater comprehension. Extensive graphics may be selected and programmed to reach beyond the mere support of narrative descriptions.

The outline advances into the next organizational step, called the *story board*. Like the card board technique, the story board is a tool that can be used to organize the visual story. This step allows the designer to place graphic elements into a position that will have the greatest impact upon an audience. The importance of each 35-mm slide, enlarged photograph, or moving scene can be properly weighed on the board for the continuity of the report's visual theme. (Figures 3-1 and 3-2)

THE 35-MM SLIDE PRESENTATION

This type of report is the most popular with many consulting organizations. It offers a great range of applicability for the presentation of design facts and information and is usually well received by audiences. Four important criteria are necessary for the writer's repertoir:

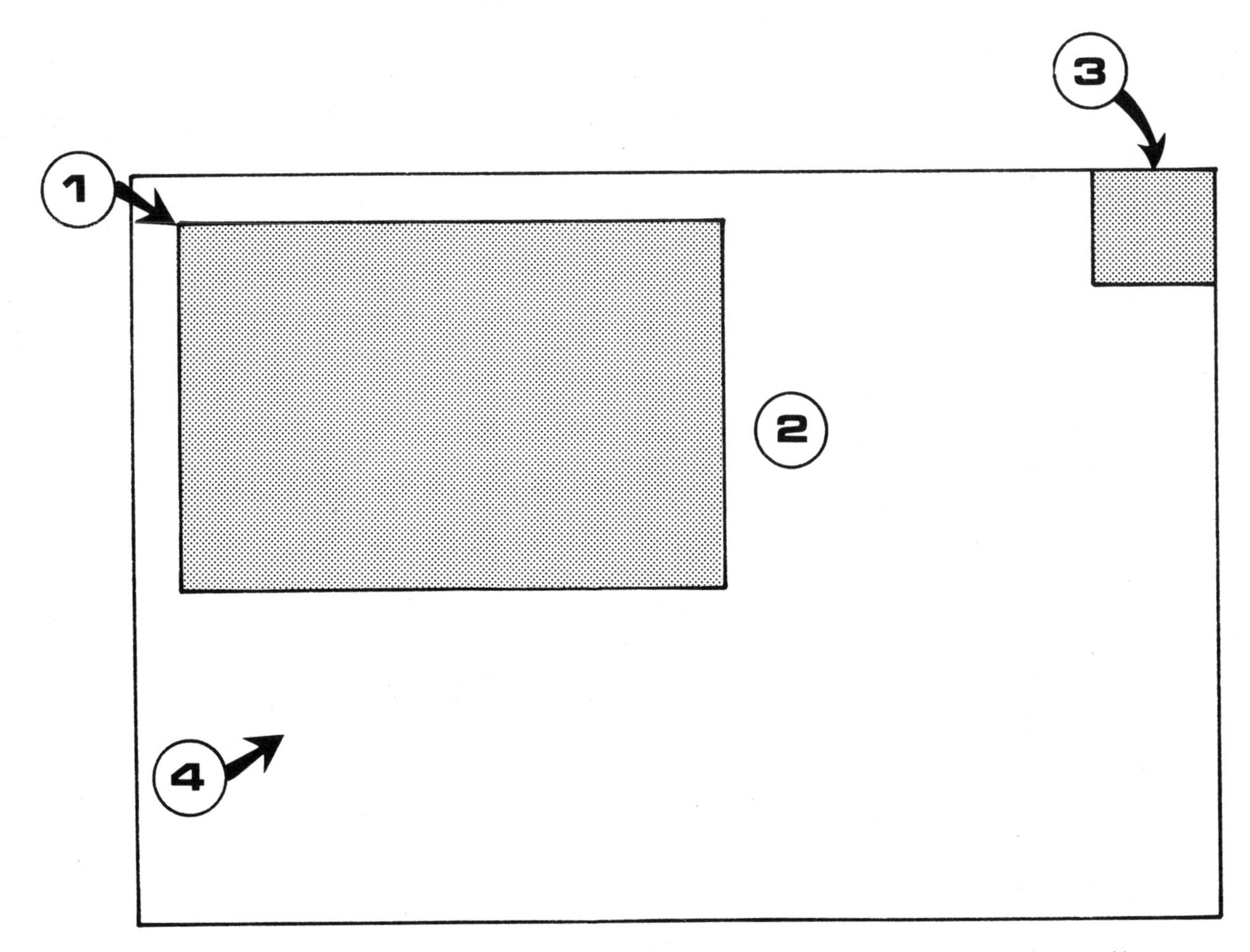

Figure 3-1. The story board card should be at least 4 × 8 inches in size. Area *1* can be used to sketch the intended slide scene. Area *2* can serve as the space for production notes, while area *3* can designate the slide number. The audio script for the slide should be written in area *4*

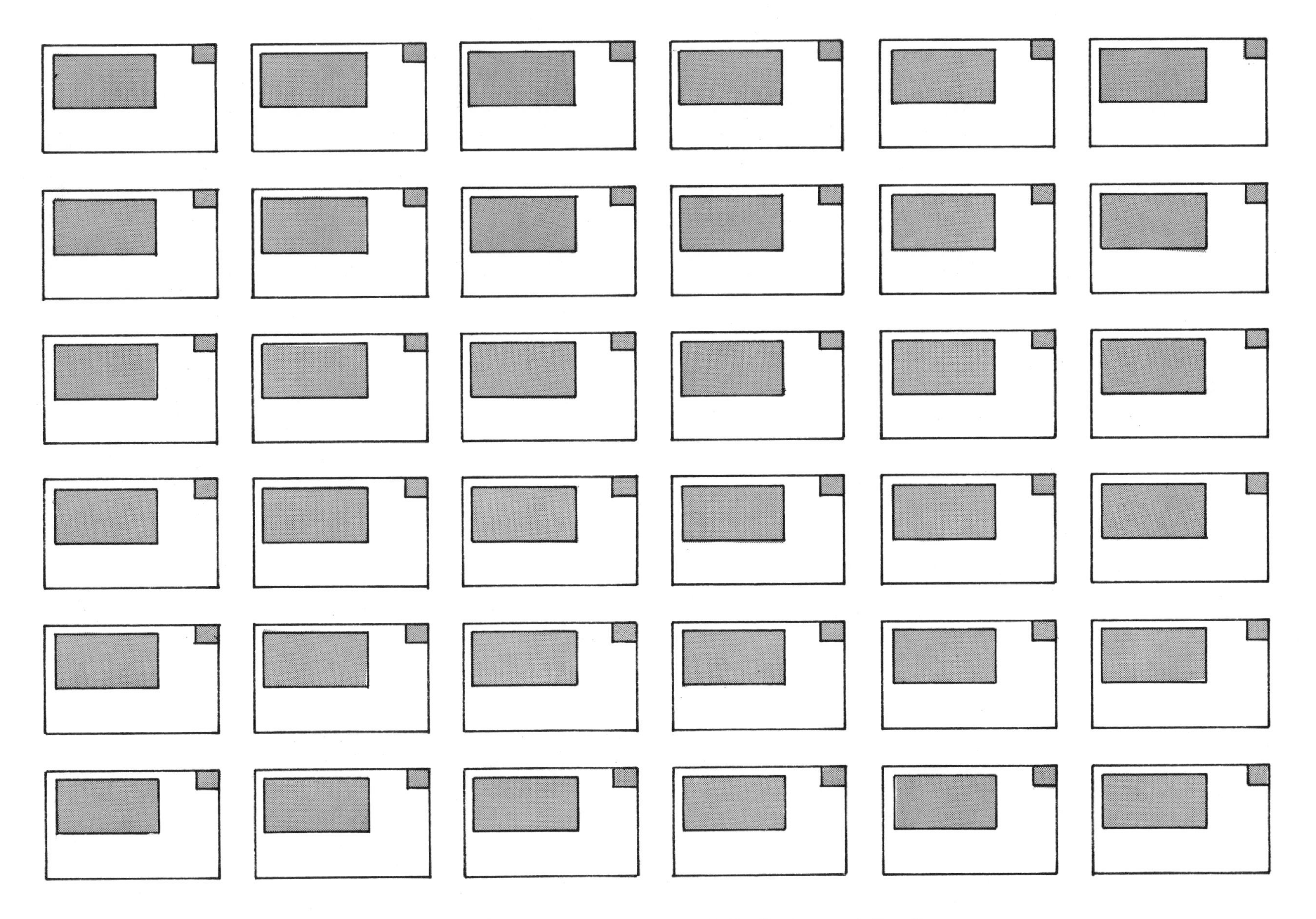

Figure 3-2. The story board should be set up to simulate the final presentation. The sequence of the slides should move from upper left to lower right.

1. *The mounting of the slides.* The amount of graphic exposure of the subject is the key to visual impact. In some cases, it is best to show only part of a photographed subject. This helps to heighten interest in the topic and to maintain that interest throughout the report. (Figure 3-3)
2. *The number of screens for the presentation.* Some reports will have greater impact upon an audience if more than one screen is used. Numerous variations are possible, but the most popular are the single, double, or multiple screen shows. The size of the viewing area is the key. Be sure each screen can be seen at the proper angle from every seat in the audience. (Figure 3-4)
3. *The audio support.* It is important that the program have some type of audio support for the visuals. Music or verbal narration will provide continuity to the report and augment the information presented by the visuals. If music is used, it should not conflict with the subject and should "move" the slides as they are changed. Verbal support should not be continuous; it should be offered only when necessary to emphasize a particular point.
4. *Supportive equipment.* The type of equipment selected to "run" the images past the audience is important. Quick-change, repetitive graphics will help sell some points, while slow phasing is best for others. Most A/V rental firms keep a complete line of supportive equipment in stock. The following accessories are among the more popular ones used for design presentation:
   a. *Synchronized recorder.* A recording of a story line can be keyed to the presentation text. The changing of the slides is controlled by a magnetic "cue" on the tape.
   b. *Dissolve control.* This attachment allows each slide to fade in and fade out without the need for a blank screen.
   c. *Programmer.* A presentation can be made with up to nine projectors at once with the assistance of a program control.
   d. *Special lens.* With a short or long focal length attachment, one slide image can be altered for different effects.
   e. *Rear-screen projection.* By carefully adjusting the slides, the projection can be completed from the rear of the screen to prevent conflicts with a verbal presentation.

THE DESIGN EXHIBIT

This form of report is used most often for the general, public audience. Bank lobbies, corporate showrooms, or professional meetings are the ideal locations for these compositions. The number of visuals selected, however, should be fewer than for other A/V forms and they must be used with care and discretion.

The viewer must approach the exhibit from a specific direction to learn the important issues of the report. Presentation information must be brief and factual but maintain the theme of the story. (Figures 3-5 and 3-6)

THE MOTION PICTURE

The range of possibilities of this report form is endless: from a simple videotape played on a client's television to a 16-mm film shown in a large auditorium. New construction techniques, design procedures, or even postconstruction evaluations can be recorded and employed for extensive educational purposes. The most practical method for developing this report form, however, is to complete the detailed story outline and then employ the services of a film producer. In-house production by the average design firm will create too many complications for efficient composition.

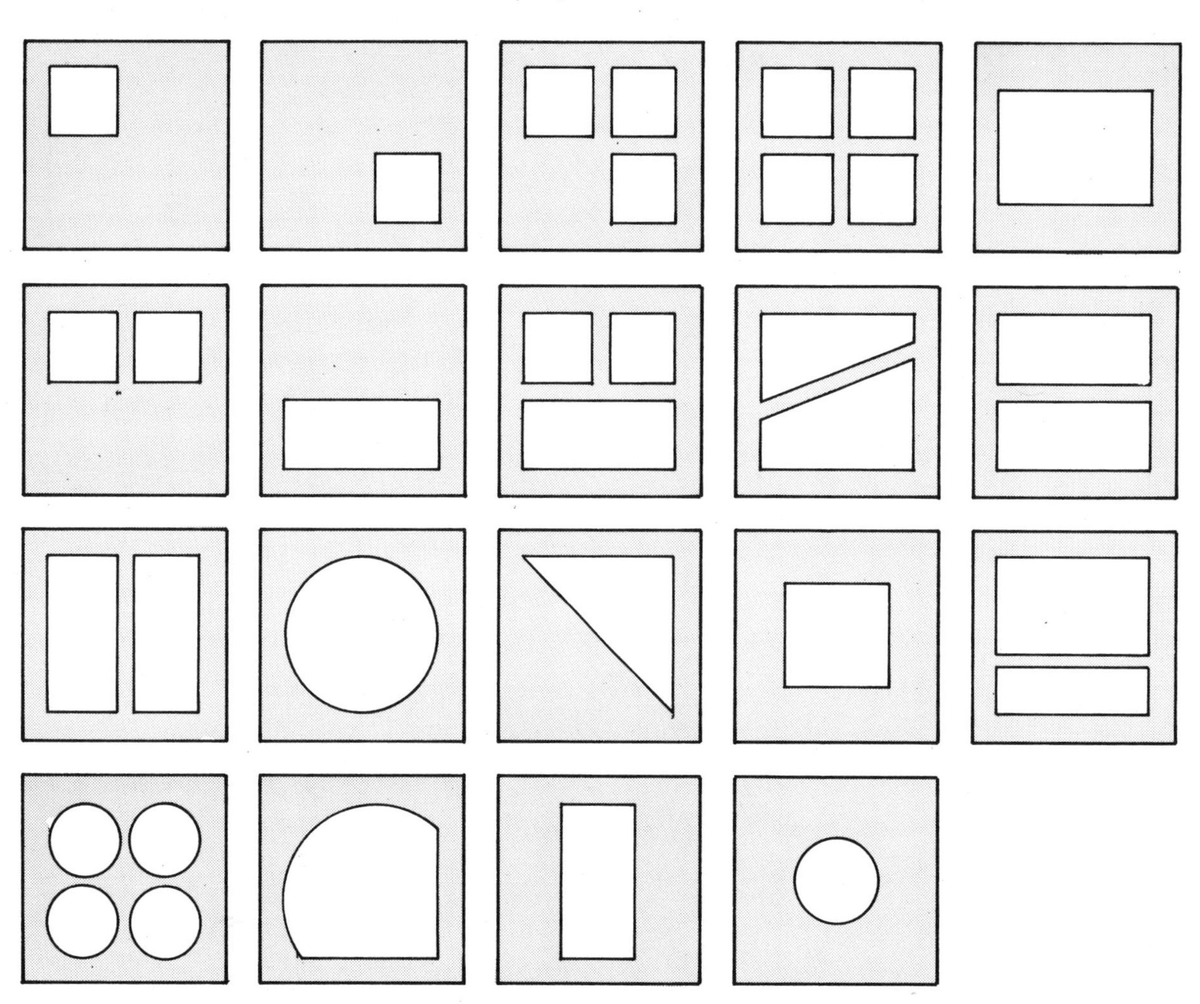

Figure 3-3. For certain effects, a designer should consider variations in the standard slide mounting.

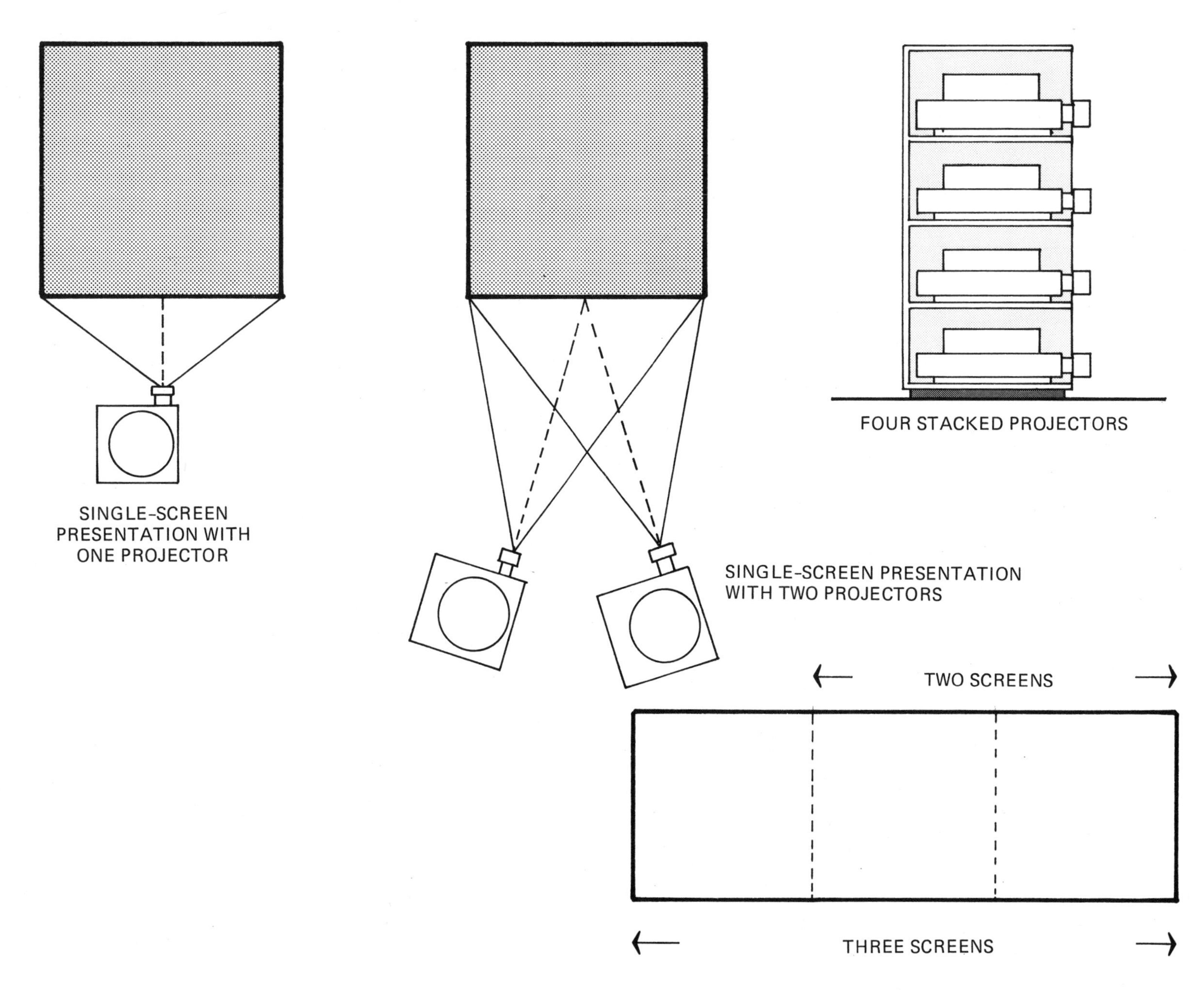

Figure 3-4. A single-screen presentation can be completed with one, two (side-by-side), or as many as four (stacked) projectors. If more than one image area is needed, the appearance should resemble one wide image.

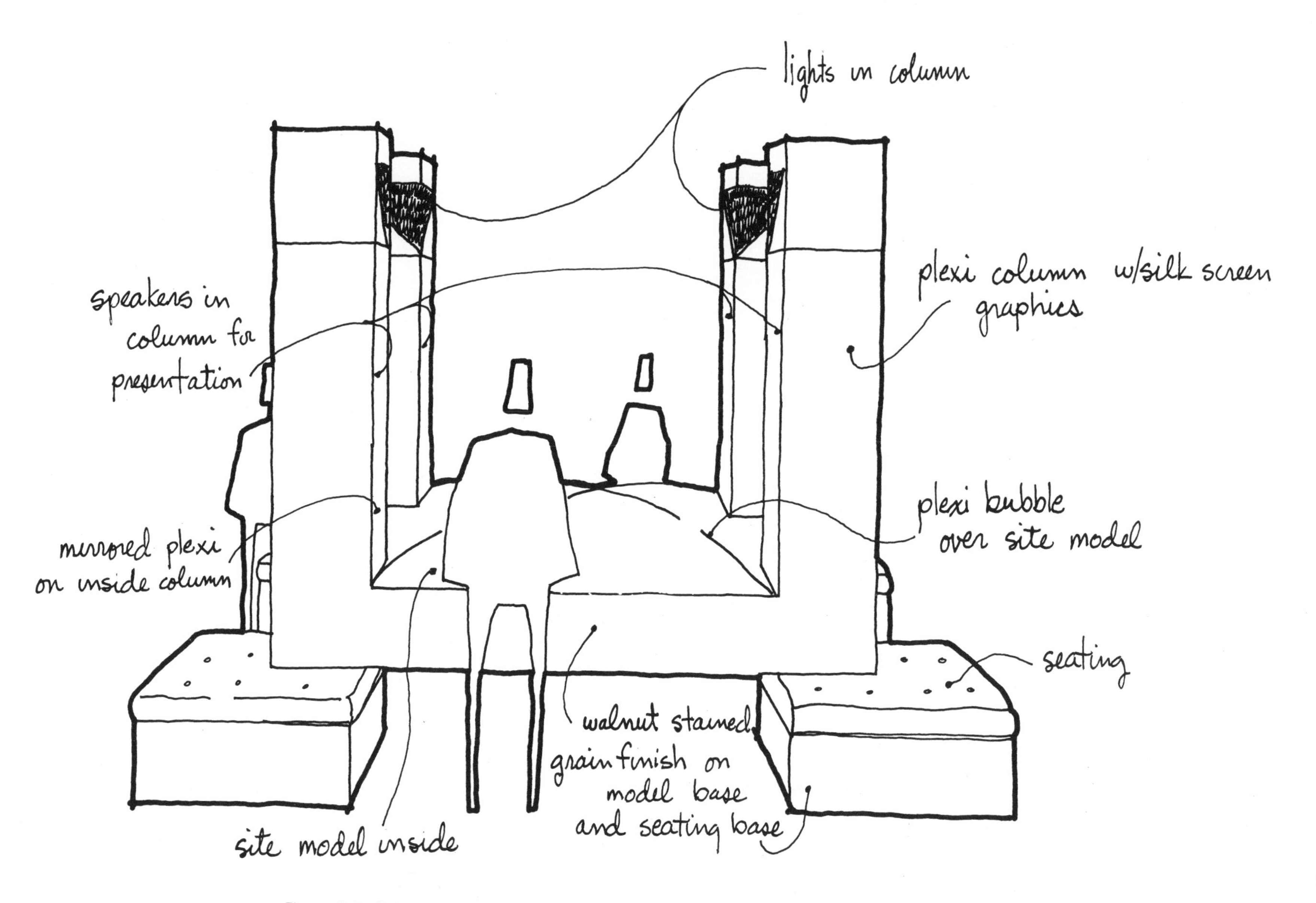

Figure 3-5. This exhibit was a report to a general audience on the final development plan of a large housing project.

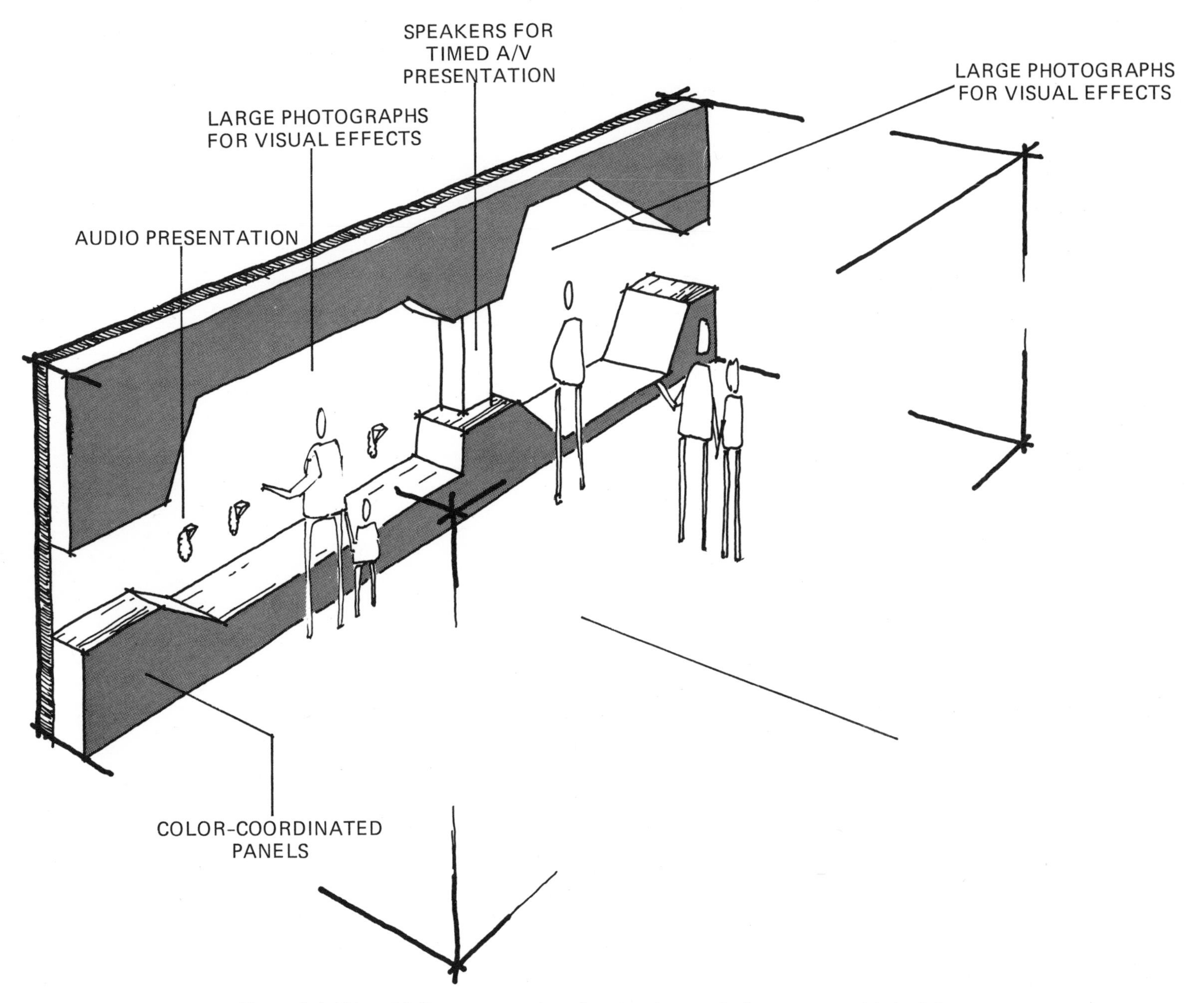

Figure 3-6. This exhibit was a report on the development of a campus-wide building program for a network of community colleges. It was placed in a small mobile trailer and "presented" at more than one location to maximize public awareness.

Appendix 1. Folding

When office brochures or large page layouts are contemplated for the design report, it is important to know how they will appear in the final composition. Large maps and detailed line illustrations are an integral part of a report's character, and it is costly when time is expended on a product that does not "fit" well within the final document. It is best to know what is needed and how it should be composed before the drafting is begun. This will save both time and money in the long run.

The following are standard paper folds for design brochures and architectural mapping elements.

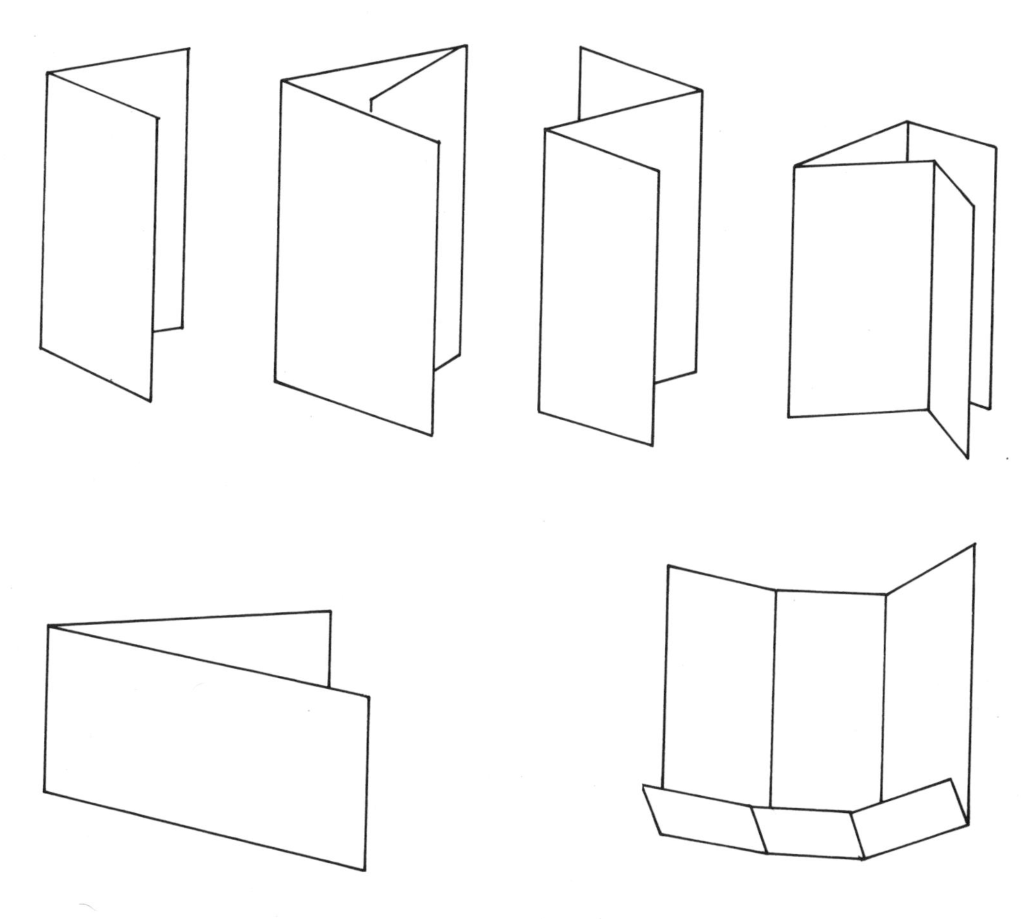

Figure A1-1.

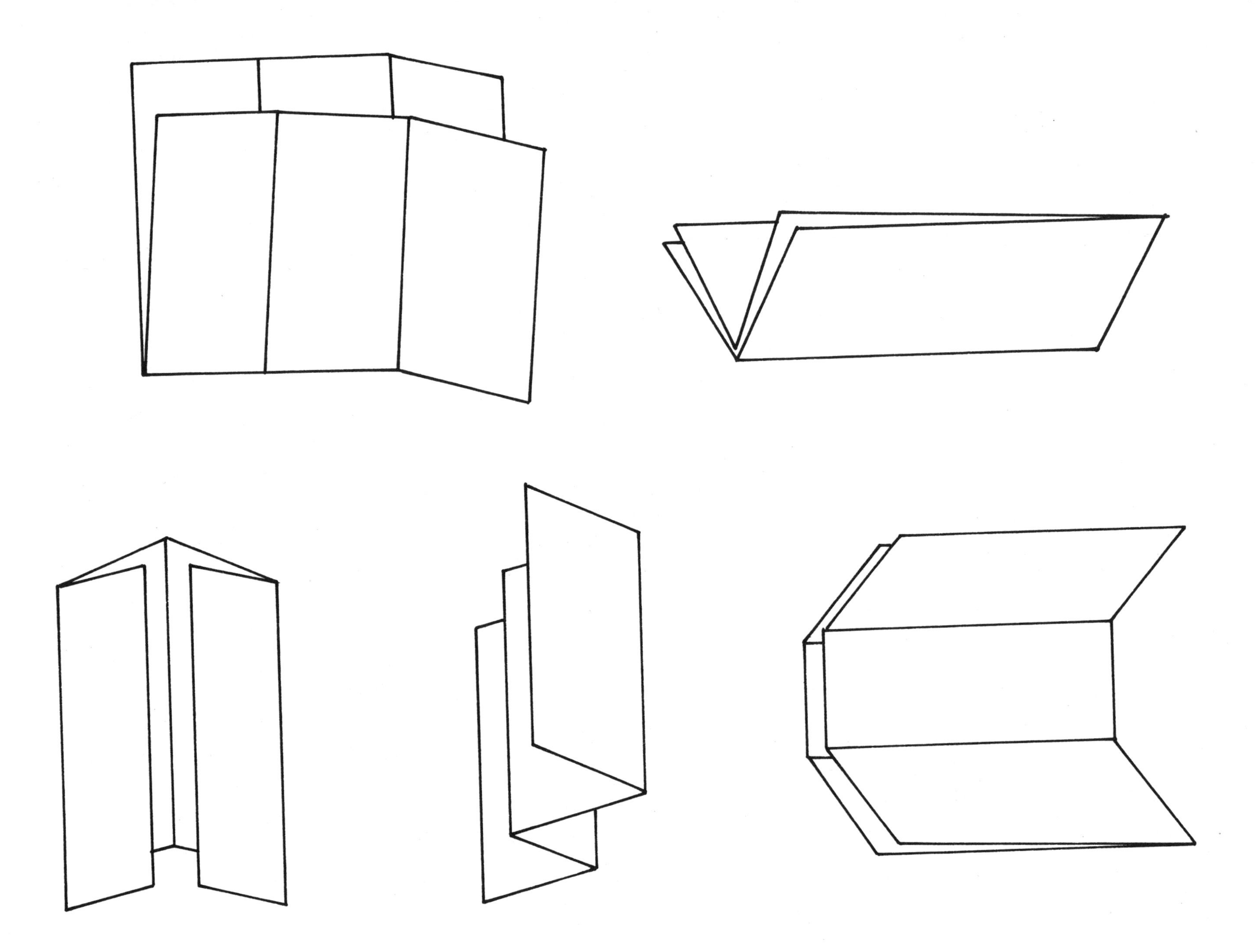

Figure A1-2.

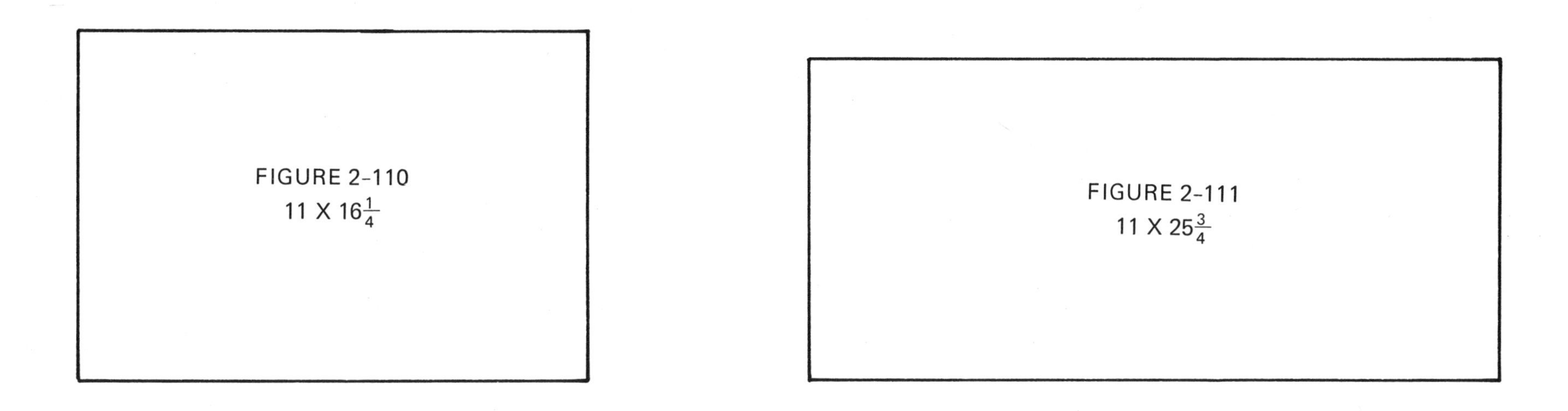

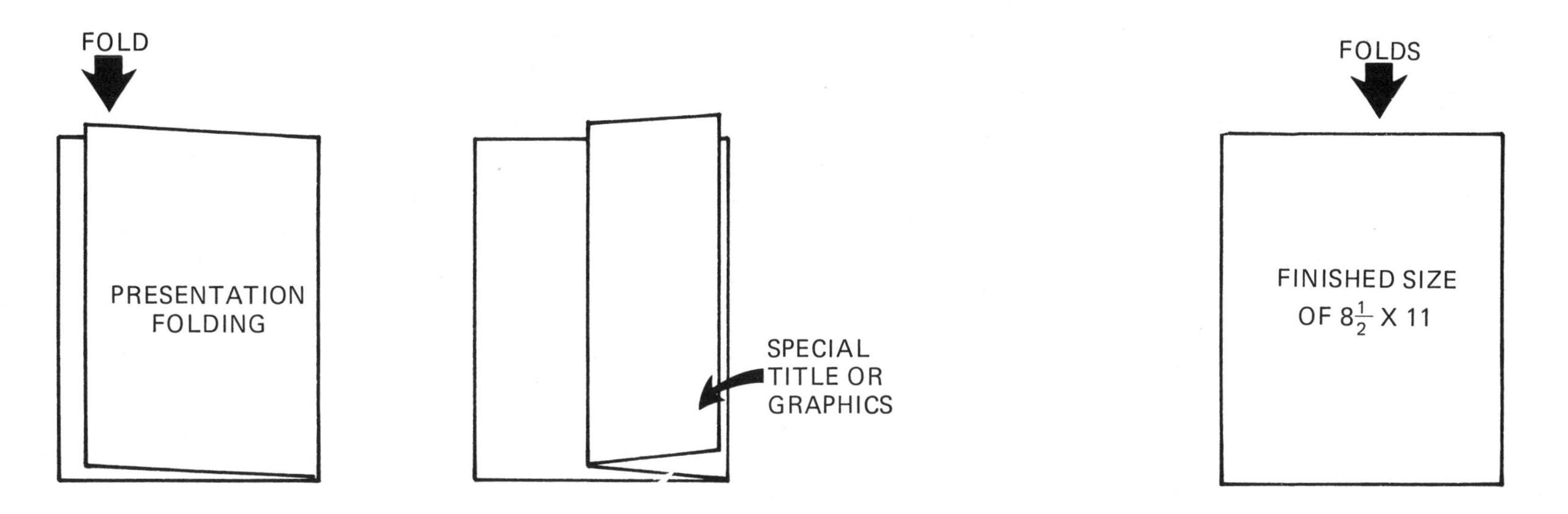

Figure A1-3. The page size on the left is 11 × 16¼. It was folded once to fit into the finished document. If it had a special title or graphics, it could have been folded once again to display these elements. The page size on the right is 11 × 25¾. It was folded twice to a finished size of 8½ × 11 inches.

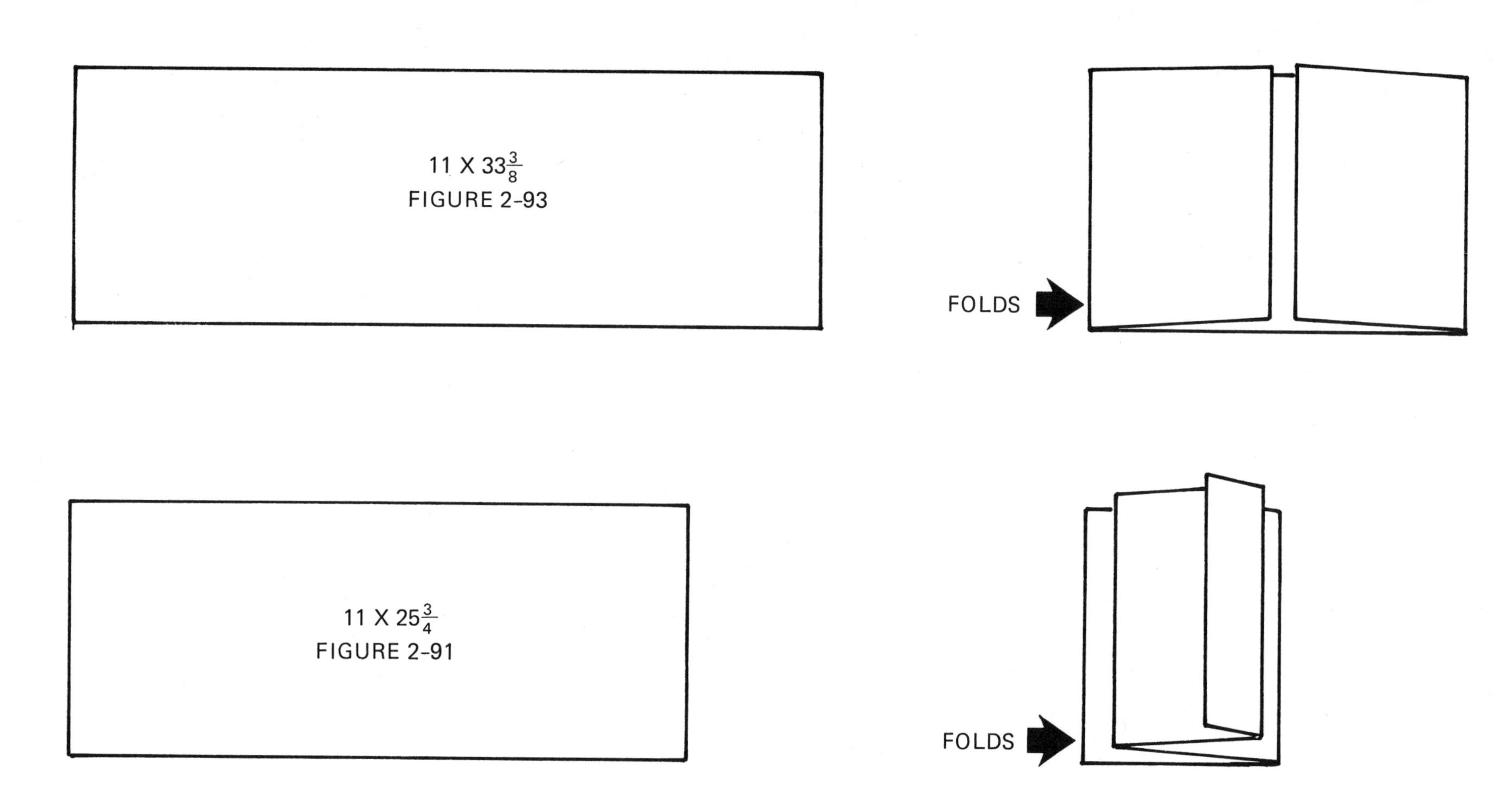

Figure A1-4. The upper page is 11 × 33⅜. It was folded three times and bound in the center of the report by a saddle-stitch. The lower page is 11 × 25¾ and is represented by Figure 2-94. It was folded three times.

Appendix 2. Binding

There are five principal methods for binding the pages of a report together:

1. *Saddle-stitch binding.* Two or more staples are inserted through the center fold of the report. Each page in the front is therefore continuous with one in the back. If full color is used, it may have to appear both in the front and the back. This method is relatively inexpensive and will allow the report to open flat, as with most magazines. However, if a large number of pages is anticipated, other forms of binding should be considered: vital information at the edges of the center pages may be cut away when the document is trimmed. Saddle-stitch binding is usually restricted to the use of only one type of paper stock for the text. (Figure A2-1 and A2-2)
2. *Perfect binding.* Adhesive material is used instead of staples to hold the binding edge together. Each page opens flat, so little space is lost in composition. Different paper stocks can be used throughout the text without extensive additional costs. Its main weakness is the chemistry of the adhesives: when subjected to extremes of heat, some glues break down and release the pages. (Figure A2-3 and A2-4)
3. *Plastic binding.* This is the most common of all the design report bindings. Many firms have machines that will punch and bind papers as needed, and the process can be less costly if completed in-house. The margin for the bound edge should be wide enough to allow for the plastic inserts. Numerous color combinations are possible with this technique. Paper stock is usually not a restriction, and the final document will lay flat on a drawing board or table. (Figure A2-5)
4. *Velo binding.* This is a relatively new type of binding for the design report and incorporates the use of a series of small holes with a plastic "clamp." It offers many advantages for smaller reports or proposals, and can be completed on an in-house machine.
5. *Ring binding.* The standard three-ring binder is commonly used when report materials reach rather large proportions. Pages are interchangeable and the final document can be expanded or updated easily. Tear and loss of individual pages seem to be the most troublesome features. (Figure A2-6)

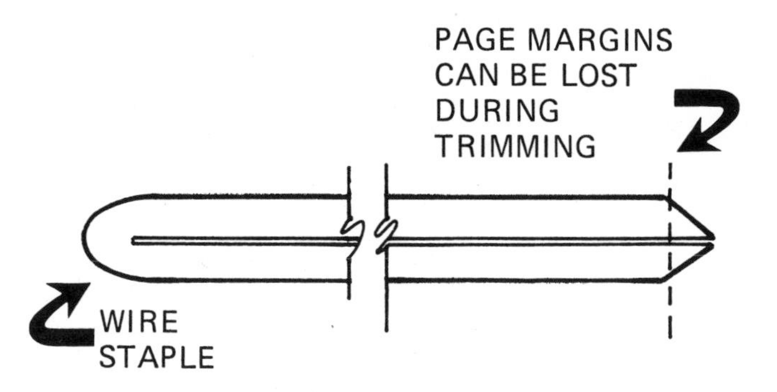

Saddle-stitch binding

Figure A2-1.

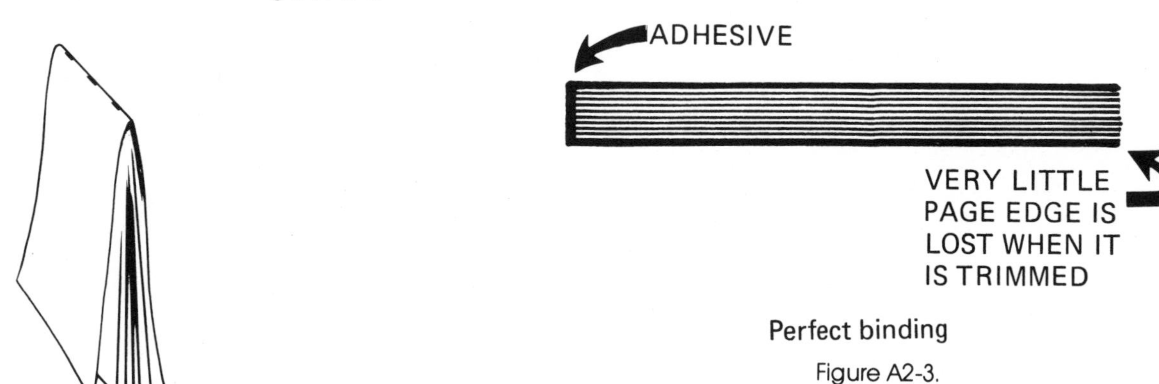

Saddle-stitch binding

Figure A2-2.

Perfect binding

Figure A2-3.

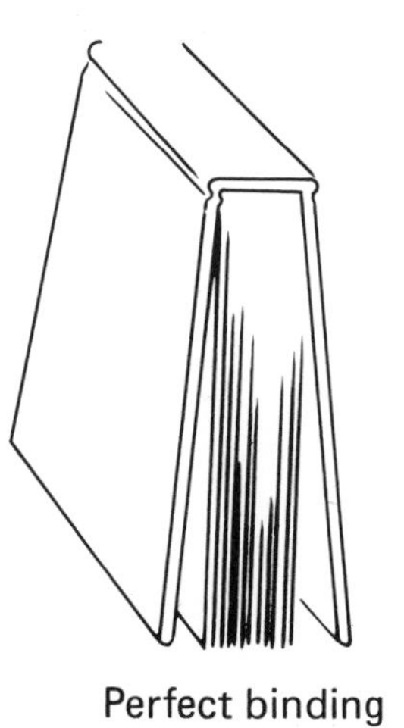

Perfect binding

Figure A2-4.

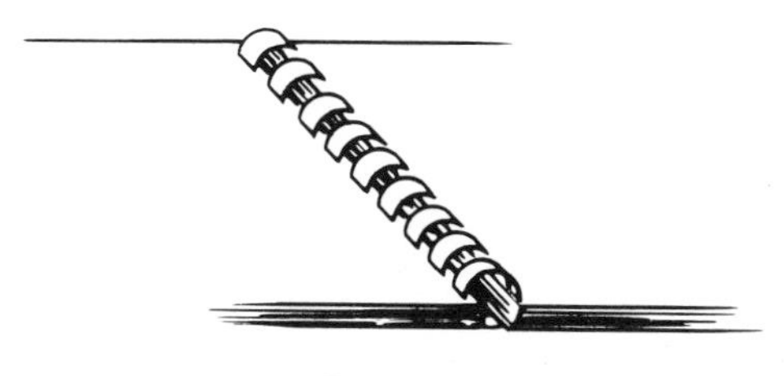

Plastic binding

Figure A2-5.

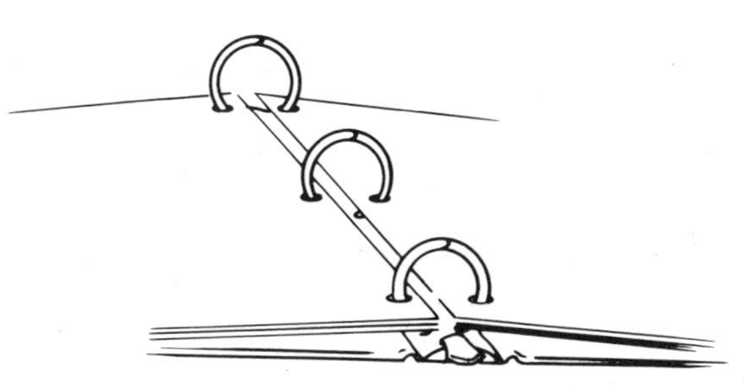

Ring binding

Figure A2-6.

Appendix 3. Paper and Reduction

PAPER

Most design and planning organizations use standard paper when preparing their design reports. Little emphasis is placed on this issue because most writers assume there are few available options. The contrary is true. Paper is available in a variety of different sizes as well as weights, grades, and colors. It is best to consult with a local printer to determine what is currently available.

Four basic page sizes are used for reports: *small, standard, medium,* and *large.* Standard is the most common, but the choice should be based upon the desired visual impact the document is to have on the reader. For example, some larger or elongated sheets are often more attractive to readers and readability will also be enhanced with their use. (Figures A3-1 and A3-2)

Paper is also available in different weights. Its basis weight is calculated by the weight of one ream of 500 sheets. For example, in 60-pound paper, one ream of 500 sheets will weigh 60 pounds. Design reports should contain at least 20 pound stock to prevent tearing and the "see-thru" effect caused by lighter papers.

The grade of paper offers the designer another choice. Paper grades used most often are *book, bond, newsprint, coated card,* and *cover. Book* paper will make a more durable and attractive text stock for the design report.

REDUCTION

The reduction of detailed drawings should be carefully considered: the drawings must be readable as well as useful for the report. The following are reduction percentages for typical architectural drawings:

	Percentage of Original Size		
Sheet Size	75%	50%	25%
8½ × 11	6⅜ × 8¼	4¼ × 5½	4⅛ × 2¾
11 × 17	8¼ × 12¾	5½ × 8½	2¾ × 4¼
17 × 22	12¾ × 15	8½ × 11	4¼ × 5½
24 × 36	18 × 27	12 × 18	6 × 9
30 × 40	22½ × 30	15 × 20	7½ × 10

The standard size for lettering on architectural drawing is ¼ inch (approximately 14 to 18 point). If a reduction of more than 75 percent of the original is required for the report, it may be best to draw the illustration first, reduce it to the desired size, then apply the typeset wording. Typical handlettering, when reduced more than 75 percent of the original may be too difficult to read.

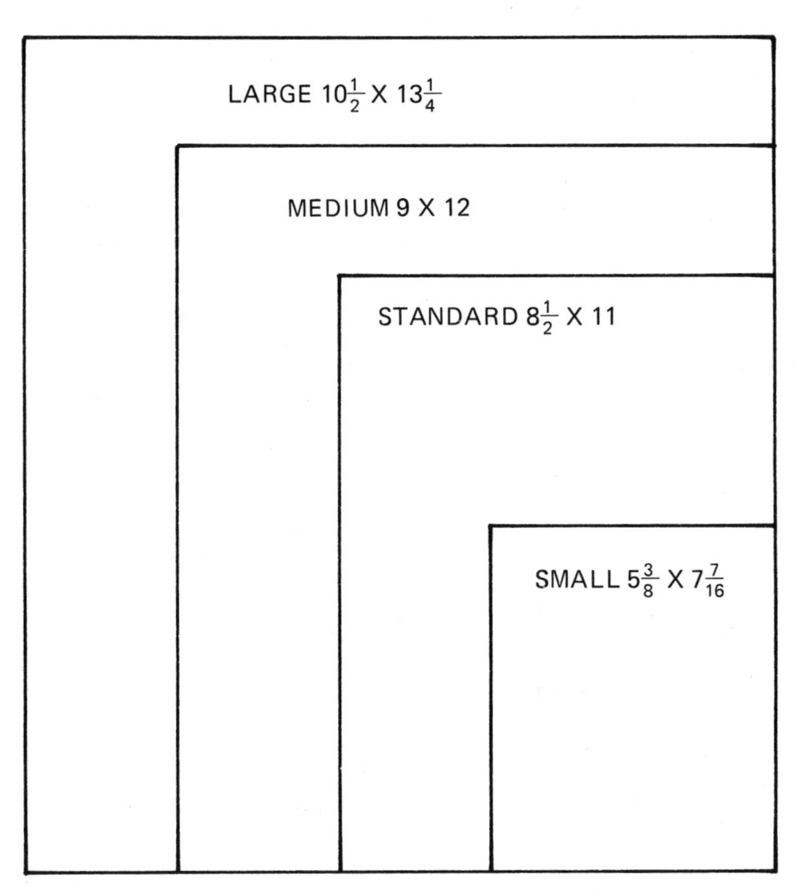

Figure A3-1.

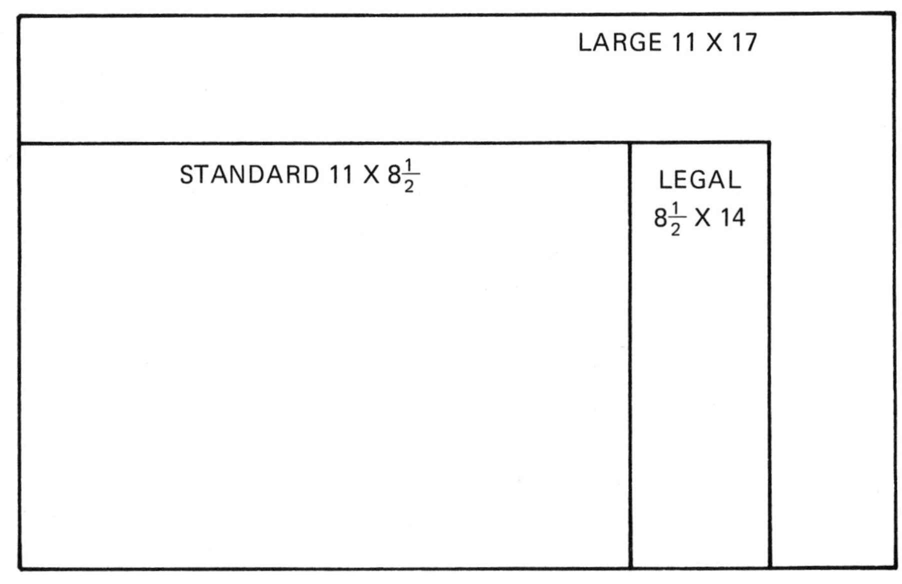

OBLONG PAGE SIZES

Figure A3-2.

Appendix 4. Slide Copy Table

Many artists and report designers use seemingly detailed and illustrative line art in their report productions. But what appears to be an elaborate street scene or captured action may actually be copied slide art. The technique is simple. A copy table is constructed to project a slide onto a glass working surface, whereupon the artist reproduces the slide image on paper. The oversize drawing is then reduced to the required dimension for the final report. Figures A4-1 through A4-3 show a typical design for a slide copy table.

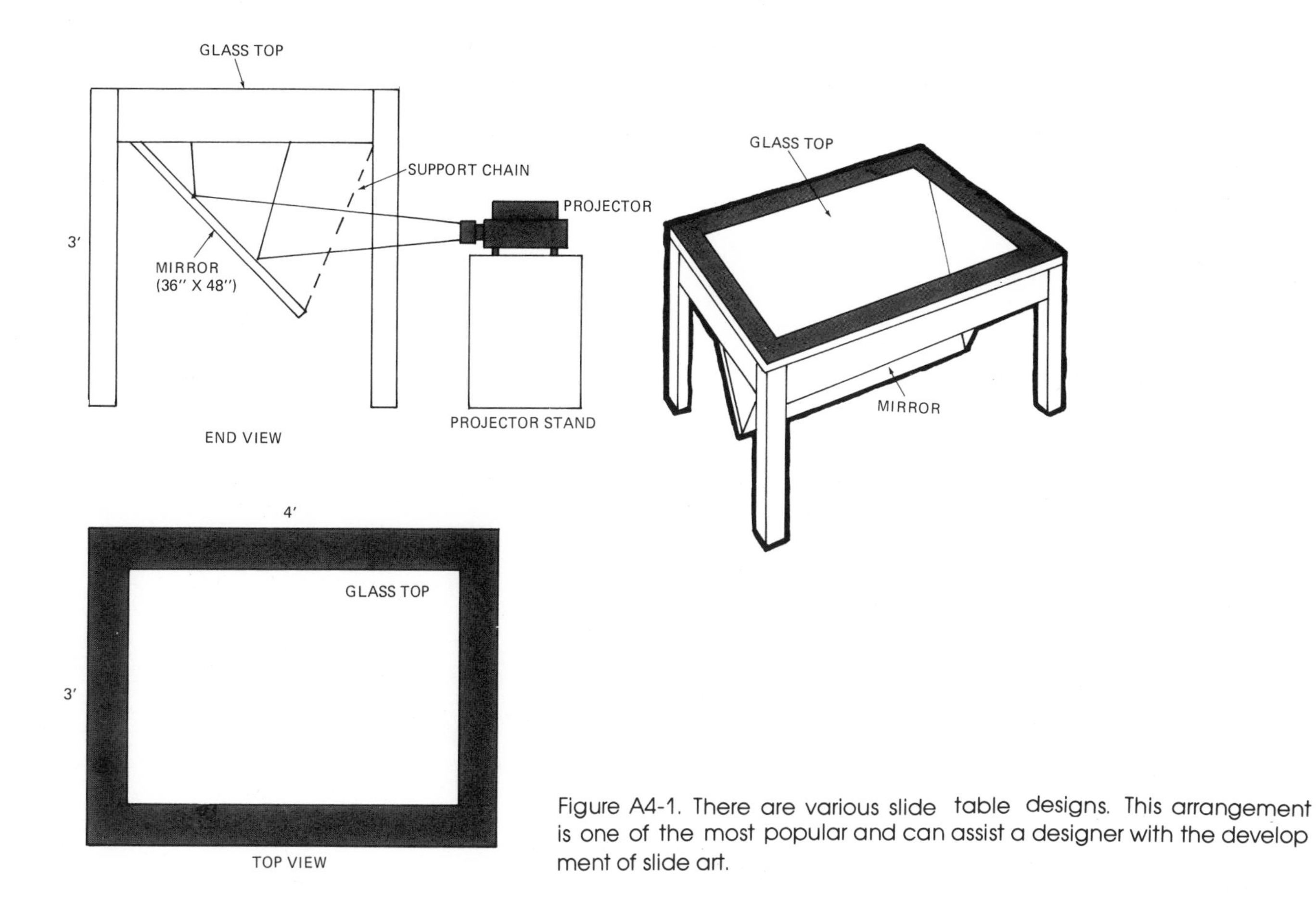

Figure A4-1. There are various slide table designs. This arrangement is one of the most popular and can assist a designer with the develop ment of slide art.

Figure A4-2.

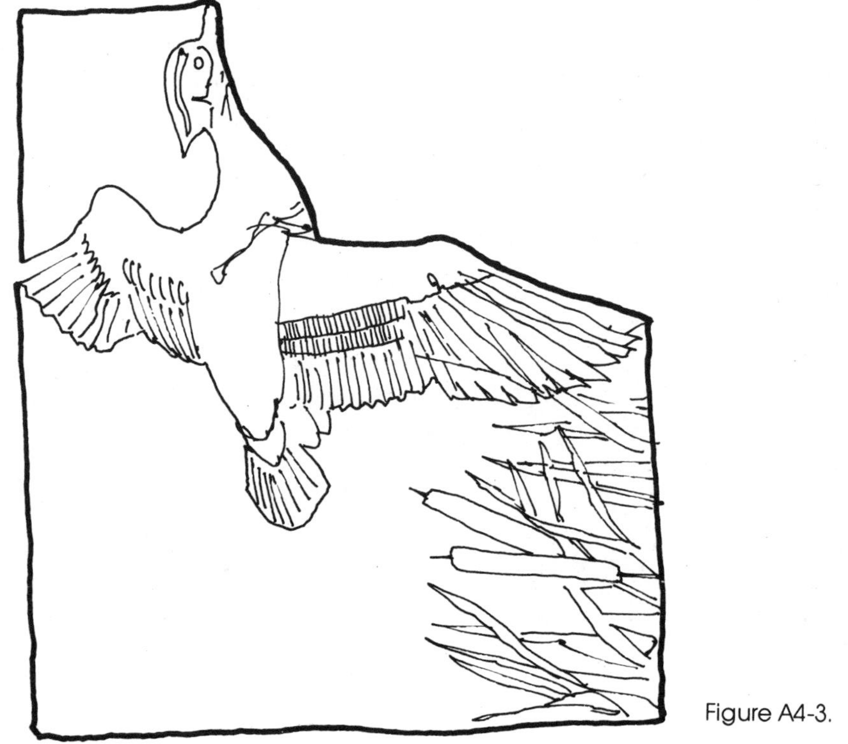

Figure A4-3.

Appendix 5. Tables and Graphs

Although most tables or graphs are not actually considered artwork or illustrative graphics, they do play a role in the production of a design report. Tables are often used to summarize research findings and assist readers in understanding complicated information. Graphs (or charts) are used more for visual impact to clarify otherwise bland or confusing statistical data.

The basic format for a *table* should be as follows:

Stub Head	**Column Head**	**Column Head**
Stub	Data	Data
Stub	Data	Data
Stub	Data	Data
Total	Sum	Sum

To develop a table for a design report:

1. For increased readability, skip at least three spaces above and below the table.
2. Place the table on the same report page as its reference in the text.
3. Use a table number for each table, and reference it in the text and at the beginning of the report.
4. Footnote the source of the information presented in the table if necessary.

There are four types of *graphs* that can be used in a design report: the line graph, the matrix, the bar graph, and the circle graph. The *line graph* is best used to describe one or more variables that occur over a period of time. The vertical axis can be used to illustrate the quantity of measurement, with the horizontal axis showing the period of time. (Figure A5-1)

The *matrix* is useful for illustrating the comparison between variables. The axis of each variable should converge to a common point. (Figure A5-2)

The *bar graph* can represent one or more variables over a period of time. It also requires a vertical and a horizontal axis, with the quality of the data represented by each bar. (Figure A5-3)

Circle graphs are used to represent percentages of a given factor. Each division, or percentage, is represented by a portion of the cricle. (Figure A5-4)

Chart I
Population Trends from 1920-1970

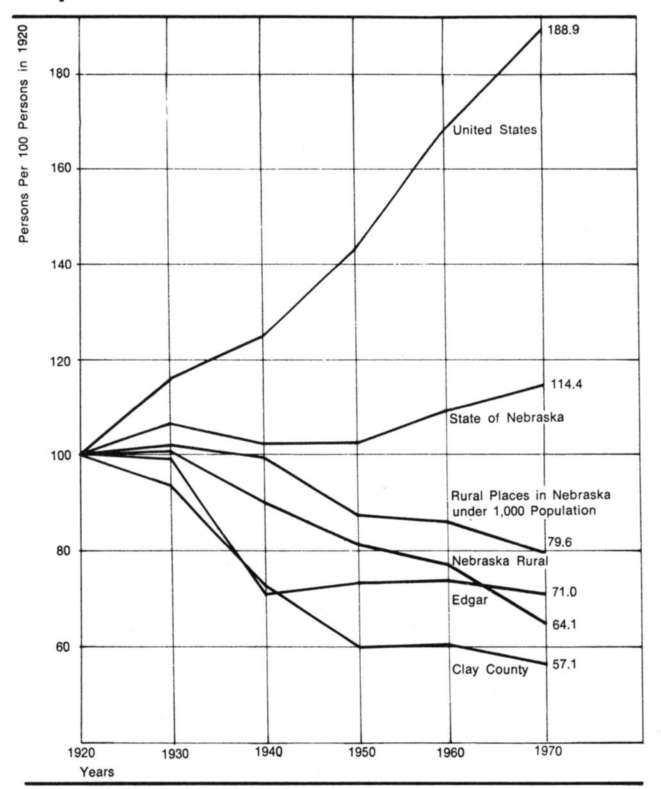

Population Composition

Figure A5-1. Line graph. Remember: the "Y" axis should be zero; tick marks should be clearly marked; and the space between the tick marks should be the same.

CALIFORNIA

	STATE FUNDED BLDG. AND FACILITIES	PRIVATELY FUNDED BLDG. AND FACILITIES	STATE AND COMMON SCHOOLS
TRANSIT SYSTEMS			
special vehicles			
special landing areas			
SITE DEVELOPMENT			
grading	●	●	●
walks	●	●	●
parking spaces	●	●	●
ramp gradients	●	●	●
landing areas	●	●	●
surfacing	●	●	●
signing	●	●	●
lighting	●	●	●
plant materials			
maintenance			
EXTERIOR BLDG. ACCESS			
entrance threshold	●	●	●
access orientation	●	●	●
surfacing	●	●	●
signing	●	●	●
INTERIOR BLDG. ACCESS			
doors and doorways	●	●	●
floors	●	●	●
thresholds	●	●	●
hardware requirements	●	●	●

	STATE FUNDED BLDG. AND FACILITIES	PRIVATELY FUNDED BLDG. AND FACILITIES	STATE AND COMMON SCHOOLS
access orientation	●	●	●
elevators	●	●	●
places of assembly	●	●	●
controls	●	●	●
COMFORT FACILITIES			
toilet rooms	●	●	●
water closet stall	●	●	●
lavatory	●	●	●
urinal	●	●	●
accessories	●	●	●
drinking fountain	●	●	●
telephone	●	●	●
GRADE CHANGES			
curbs	●	●	●
ramps with handrails	●	●	●
expansion units			
stairways	●	●	●
hazard warnings	●	●	●
RECREATION AREAS			
picnic facilities			
camping facilities			
therapeutic playgrounds			
general outdoor use			
surfacing			

Figure A5-2. Matrix graph.

APPLICATION: The California state statutes apply to all buildings and facilities constructed by the state of any of its political subdivisions. The statutes also apply to auditoriums, hospitals, theaters, restaurants, hotels, motels, stadiums, and convention centers constructed by private or state funds.

SPECIFICATIONS AND REGULATIONS: Adapts the American Standards Association specifications A 117.1 - 1961 for making buildings and facilities accessible to and usable by the physically handicapped.

ESCAPE OR WAIVER CLAUSES: The statutes allow for the waiving of standards if undue hardships or just reasons are given.

SPECIAL CONSIDERATIONS: There are no special considerations which may further enhance a barrier free environment.

ENFORCEMENT: The state architect is responsible for enforcement of the law in regards to building design.

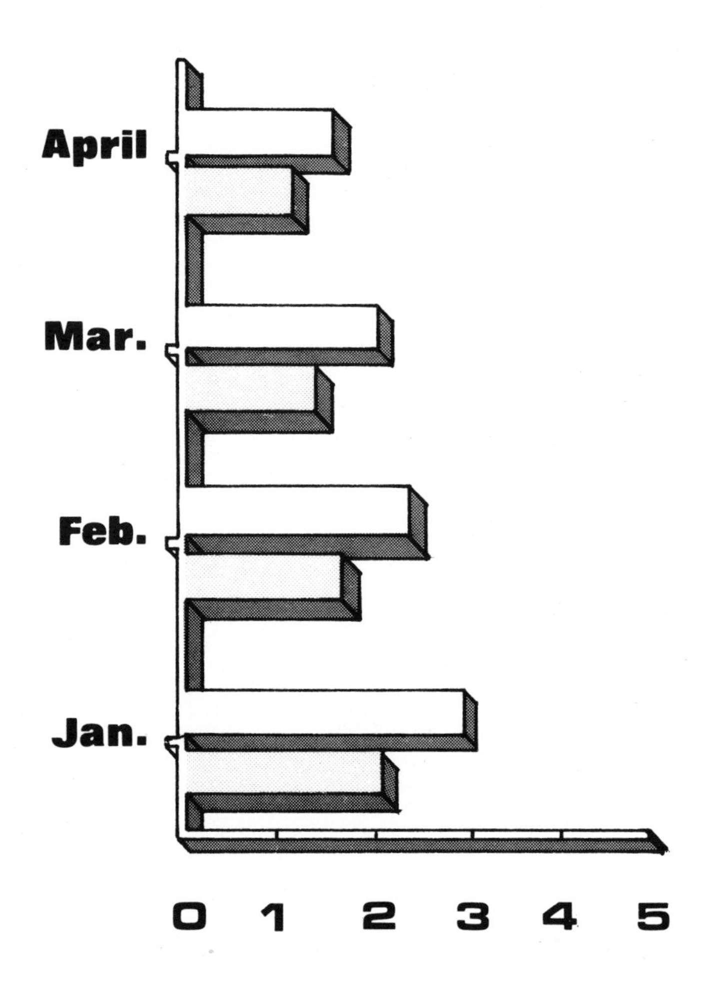

Figure A5-3. Bar graph. Remember: the width of the bars should be equal; the space between the bars should be equal; label the bars clearly; and the "Y" axis should begin with zero.

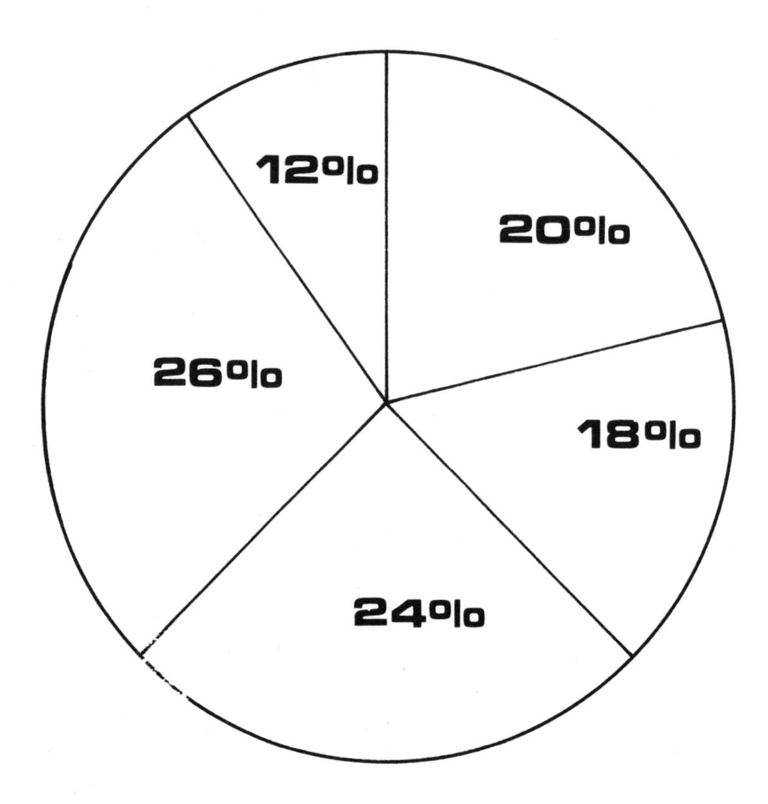

Figure A5-4. Circle graph. Remember: the first slice of the circle should be drawn at 12 o'clock; each segment should be identified; and percentages should be shown in each segment.

PRODUCTION CONTROL FORM

Structure

Front Matter

- ☐ Front Cover ____________
- ☐ Title Page ____________
- ☐ Preface/Forward ____________
- ☐ Acknowledgements ____________
- ☐ List of Illustrations ____________

Main Body

- ☐ Introduction ____________
- ☐ Summary ____________
- ☐ Recommendations

Back Matter

- ☐ Appendix ____________
- ☐ Glossary ____________
- ☐ List of Refefences ____________
- ☐ Bibliography ____________
- ☐ Index ____________
- ☐ Back Cover ____________

Audience

- ☐ Technical ____________
- ☐ General ____________
- ☐ Management ____________

Page Composition

- ☐ Type Style ____________
- ☐ Margin ____________

Support Material

- ☐ Photos ____________
- ☐ Line Art ____________
- ☐ Symbolic Art ____________

Paper

- ☐ Stock ____________
- ☐ Size ____________

Reductions

Original Size	Finished Size

Preliminary Draft

CHECK:

- ☐ Sentence Structure ____________
- ☐ Troublesome Words ____________
- ☐ Word Usage ____________
- ☐ Caps ____________

Paragraph Composition

- ☐ Length ____________
- ☐ Unity ____________
- ☐ Flow ____________

Proofread ____________

Readability ____________

Final Typing/Printing

____

____

____

____

____

____

Binding ____________

Folds ____________

ISSUED TO CLIENT ____________

Glossary

appendix—the supplemental information attached to the end of a design report.

artwork—material such as line drawings, photographs, tables, graphs, charts, etc., used to support the written text of a design report.

audio—the tape recording used to support the story line of an AV report.

back matter—the part of the final report containing the appendix, the glossary, references, bibliography, index, and the back cover.

basis weight—the weight of the paper used for a design report.

bibliography—a list of publications used to support the text of a design report.

bind—to apply the parts of the report together into one document.

bleed—a photograph or line art illustration off the edge of report page.

caption—the supportive description for a photograph or art used in a design report.

clause—group of related words with a subject and a verb.

collate—to arrange the pages of a design report in a specific order.

column—a part of a report page for the placement of text or art.

complex sentence—a sentence containing one independent clause and at least one dependent clause.

composition—the arrangement of the design report.

compound-complex sentence—a sentence containing at least two independent clauses and at least one dependent clause.

compound sentence—a sentence containing two or more independent clauses.

contoured—a type of column margin.

copy—the portion of the manuscript that is to be reproduced.

cover stock—the heaviest paper of the design report.

design exhibit—a unique form of design report used to communicate information to a general audience.

direction—the specific path of movement of a design report.

dummy—the preliminary layout of a design report.

edit—to read and correct the report manuscript.

editor's marks—a series of instructions used to note changes in a manuscript.

eye-flow—the movement of the reader's eyes across a page.

eyewash—report art that does not support the text.

flush-left margin—a type of column margin.

flush-right margin—a type of column margin.

folding—the adjustment of a report page to meet a specific visual requirement.

foldout—a special treatment of art, usually a map.

format—the arrangement of the report.

front matter—the cover, title page, preface, acknowledgments, table of contents, and list of illustrations.

galley—a reproduction of pages set in type for proofreading.

general audience—a group of readers from various backgrounds.

glossary—explanation of terms used in a report text.

grain—the textural appearance of a paper stock.

grid—horizontal and vertical lines used in paste-up.

gutter—the inside margin of a report page.

half-tone—a black and white photograph.

heading—lines of type introducing the reader to a text chapter or section.

index—a list of items in a printed work that provides page numbers for reference.

introduction—a short background statement.

justified margin—a type of column margin.

lithography—a printing process.

logotype—a single piece of type.

outline—the order of appearance of material in a design report.

overlay—a part of a series of mapping components of a design report.

perfect binding—a type of report binding using an adhesive substance to hold the pages together.

PMT—a copy of an artwork used for a design report.

polishing—determining the readability of a design report.

readability—the ability of a written text to be understood by a target audience.

readability index—a measuring system for determining the readability of report material.

reference list—a listing of sources of information for a design report.

rough draft—the first typed copy of design report.

runaround—a type of column margin.

saddle-stitch—a type of report binding technique using wire staples.

section—a specific portion of a manuscript.

simple sentence—a sentence containing only one independent clause.

story board—

subheading—the subject of a series of paragraphs.

subsection—the subject of a specific paragraph.

summary—the conclusions or findings of a report.

symbolic art—visual materials used instead of lines of type.

technical audience—a group of persons with specific knowledge of a report contents.

title—the name of the design report

typography—the type style and its use in a design report.

References

Ballinger, Raymond A. *Layout and Graphic Design.* New York: Van Nostrand Reinhold Company, 1972.

Burden, Ernest. *Visual Presentation.* New York: McGraw-Hill Book Company, 1977.

Huey, E.B. *The Psychology and Pedagogy of Reading.* New York: McMillan & Co., 1908; and Cambridge, Mass.: M.I.T. Press, 1968.

Hurlburt, Allen. *Publication Design.* New York: Van Nostrand Reinhold Company, 1971.

Lewis, Phillip V., and William H. Baker. *Business Report Writing.* Columbus, Ohio: Grid Incorporated, 1978.

McCullo, Marion. *Proofreader's Manual.* New York: Richards Rosen Press, 1969.

McLean, Ruari. *Magazine Design.* London: Oxford University Press, 1969.

McNaughton, Harry H. *Proofreading and Copyediting.* New York: Hastings House Publishers, 1973.

Moyer, Ruth, Eleanor Stevens, and Ralph Switzer. *The Research and Report Handbook.* New York: John Wiley and Sons, 1981.

Phillips, Arthur H. *Handbook of Computer-Aided Composition.* New York: Marcel Dekker Incorporated, 1976.

Smith, Courtland G. *Magazine Layout.* New York: Courtland Gray Smith, 1973.

Tinker, M.A. *Legibility of Print.* Ames, Iowa: Iowa State University Press, 1963.

Tinker, M.S. "Experienced Studies in the Legibility of Print: An Annotated Bibliography. *Reading Research Quarterly.* Vol. 1, No. 4, pp. 67-118.

Van Hagan, Charles E. *Report Writers' Handbook.* New York: Dover Publications, Inc., 1961.

Wallance, John D. and I. Brewster Holding. *Guide to Writing and Style.* Columbus, Ohio: Battelle Memorial Institute, 1966.

Index

acknowledgments 29
appendix 64
A/V reports 131

back cover 65
back matter 23
bibliography 64
binding 143

card boards 4
chapter components 64
color and tone 97
complex sentence 12
compound sentence 12
cover designs 27-53

design exhibits 134-138
discussion 64

editing 15
exhibits 134-138

folding 139
format 21
foreword 29
front cover 23
front matter 23

general audience 1
glossary 64
graphs 151
Gunning Fog Index 17

horizontal reports 21

index 64
introduction 29

line art 92
list of illustrations 29
list of references 64

main body 23
major headings 4
management audience 1
manuscript page 16
margins
  centered 81, 82
  contoured 81, 82
  flush left 65, 81
  flush right 65, 82
  justified 65, 81
  runaround 81, 83
  shaped 81, 83

objectives 1
outlines 4

page composition 65
page layouts 97-130
paper 147
paragraph
  length 12
  unity 12
photo angles 92
photographs 85
point gage 83
polishing 16
preface 29
proofreader's marks 15
proofreading 15

readability 16, 17
reduction 147
report direction 21

screens 93-95
sections 4
sentence writing 11
simple sentence 12
slide copy table 149
story card 132
sub-heading 4
sub-sections 8
supportive material 85
symbolic art 92

table of contents 29
tables 151
technical audience 1
title pages 29
toning 97-98
traffic signs 3
type spacing 91
type styles 84-90

vertical reports 21

writing style 8, 19